Stuart Walmsley

CW00819253

Uni

£7-00

Cyril F. Gardiner

A First Course in Group Theory

Springer-Verlag
New York Heidelberg Berlin

Cyril F. Gardiner
University of Exeter
Department of Mathematics
North Park Road
Exeter, EX4 4QE
England

AMS Subject Classification: 20-01

Library of Congress Cataloging in Publication Data

Gardiner, Cyril F
A first course in group theory.

(Universitext)
Bibliography: p.
Includes index.
1. Groups, Theory of. I. Title.
QA171.G27 512:22 80-25160

Printed in the United States of America.

9 8 7 6 5 4 3 2 1

ISBN 0-387-90545-6 Springer-Verlag New York Heidelberg Berlin
ISBN 3-540-90545-6 Springer-Verlag Berlin Heidelberg New York

PREFACE

One of the difficulties in an introductory book is to communicate a sense of purpose. Only too easily to the beginner does the book become a sequence of definitions, concepts, and results which seem little more than curiousities leading nowhere in particular. In this book I have tried to overcome this problem by making my central aim the determination of all possible groups of orders 1 to 15, together with some study of their structure. By the time this aim is realised towards the end of the book, the reader should have acquired the basic ideas and methods of group theory. To make the book more useful to users of mathematics, in particular students of physics and chemistry, I have included some applications of permutation groups and a discussion of finite point groups. The latter are the simplest examples of groups of particular interest to scientists. They occur as symmetry groups of physical configurations such as molecules.

Many ideas are discussed mainly in the exercises and the solutions at the end of the book. However, such ideas are used rarely in the body of the book. When they are, suitable references are given. Other exercises test and reinforce the text in the usual way.

A final chapter gives some idea of the directions in which the interested reader may go after working through this book. References to help in this are listed after the outline solutions. Also in this chapter I have included the usual results on series and solvable groups required for the study of the Galois group occurring in field theory and algebraic number theory.

In my experience, the value of a book to the average student is increased considerably by the inclusion of solutions to the exercises. This is true in particular for students who, by choice or necessity, work largely on their own. For this reason, I have included outline solutions of all the exercises at the end of the book. However, the student is advised to make a determined effort to solve the problems himself (or herself) before looking at the given solutions. I have made no attempt to grade the problems. In any case one person's difficulty may be another person's triviality.

For the basic notions of sets, relations, functions, and linear algebra the reader is referred to my book 'Modern Algebra; A Natural Approach, with Applications' (Ellis Horwood Ltd.). As a general reference this will be denoted by [G] in the text. In general, references will be denoted by [n] for some number n as given in detail in the bibliography at the end of the book.

This book is based on lectures given over many years at Exeter both to specialist mathematicians and to those whose main interests lay elsewhere yet who required the usual basic ideas in group theory. The interests of the latter are served by the first four chapters of this book, though I would be delighted should they read further.

Exeter, 1980. C. F. Gardiner

CONTENTS

CHAPTER 1. FIRST IDEAS

1.1 Introduction

There is some evidence to suggest that a sense of symmetry is at least as fundamental as a sense of number, although the concept of number predates the concept of symmetry.

An implicit use of symmetry occurs in mathematics, particularly geometry, as far back as the Greeks some 2,500 years ago. But strangely, the Greeks never captured the essence of symmetry explicitly. A detailed, but informal, discussion of symmetry is given by H. Weyl in his book "Symmetry" published by the Princeton University Press in 1952. The reader is recommended to study Weyl's book alongside the present one. For a visual representation of some of the ideas that we shall discuss, see "The Graphic Work of M. C. Escher" - Escher (Pan-Ballantine 1972, Oldbourne Press 1961).

In this book we regard group theory as the conscious and explicit study of symmetry. As such the subject emerged around 1830 A.D. At that time most of the leading mathematicians of the day were engrossed in the following problem.

For any real numbers a, b, c, with $a \neq 0$, the equation $ax^2 + bx + c = 0$ has a solution formula, namely:

$$x = \frac{-b \pm \sqrt{b^2 - 4ac}}{2a} .$$

The cubic and quartic equations also have solution formulae of a similar form involving root operations on the coefficients of the equation, though more complicated and with roots other than square roots.

Problem: Is there a solution formula of this form for **any** quintic equation: $ax^5 + bx^4 + cx^3 + dx^2 + ex + c = 0$?

Around 1830, Abel answered this question in the negative. Shortly afterwards Galois gave the complete solution for equations of any degree; and more besides. His methods exploited symmetry explicitly in a way that led to the study of groups, at least of a certain type, the so-called permutation groups.

However, perhaps the most obvious way of seeing the connection between formal group theory and the intuitive idea of symmetry that it attempts to capture is by way of geometry, as follows.

Take an equilateral triangle in space and ask how you can move it so that it appears not to have moved. This is to be the measure of its symmetry. Two motions are considered distinct if and only if their 'end effects' are different. To be more precise we number the vertices 1, 2 and 3, and consider two motions to be distinct if they produce different arrangements of the vertices in space, that is different permutations of the set {1, 2, 3} , when the triangle has apparently returned to its original position.

Take L to be a line through the vertex 1 perpendicular to the side joining vertices labelled 2 and 3 (see Figure 1.1.1).

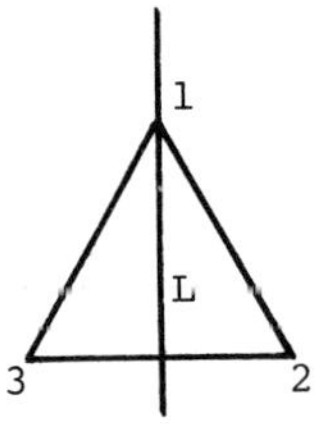

Figure 1.1.1

We consider L to be fixed in space and move the triangle about it. We consider two basic motions.

(1) A rotation about an axis perpendicular to the plane of the triangle in an anti-clockwise sense through $2\pi/3$ radians. Denote this motion by the symbol a.

(2) A rotation about the axis L through π radians. Denote this motion by the symbol b.

We write ab to denote the motion b followed by the motion a. We write b^2 to denote b followed by b, and so on.

The reader may find it convenient to cut an equilateral triangle out of cardboard in order to carry out the following manipulations. This avoids the tedium of making countless drawings.

We consider the effect of each of the following motions on the triangle in its starting position as shown in Figure 1.1.1 a, a^2, a^3, b, b^2, ab, a^2b, a^3b, ba. The results are shown in Figure 1.1.2.

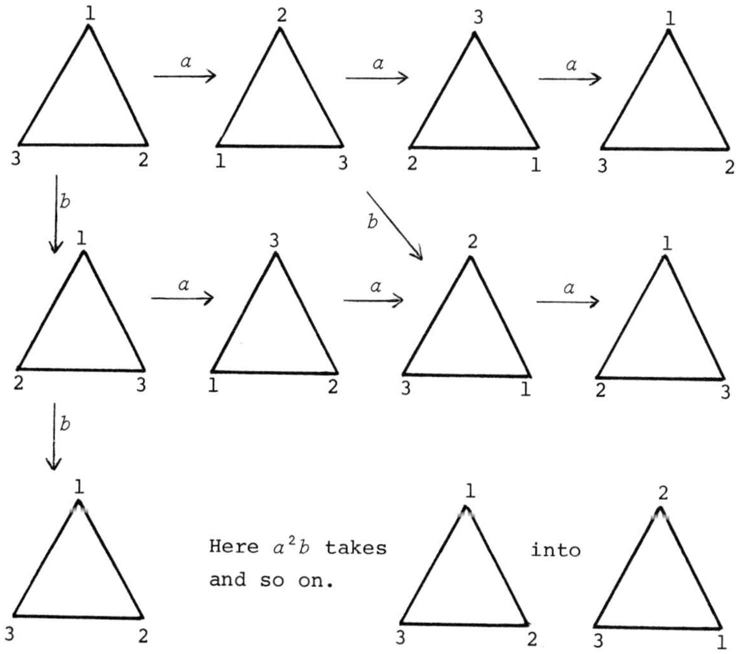
Here a^2b takes into and so on.

Figure 1.1.2

In this way we obtain the 6 possible permutations on the set {1, 2, 3} of vertices of the triangle. Since we cannot get any more permutations, we must have obtained all possible distinct motions of the triangle in space which leave it apparently fixed in space.

From a given starting position we have obtained all possible permutations on the vertices and hence all possible symmetries of the triangle, using just the motions a and b. In fact there are 6 possible motions, as measured by their effects, namely: a, a^2, b, ab, a^2b, and $a^3 = b^2 = e$, where we use e to denote the identity motion; that is the motion leaving the triangle fixed.

As is customary in modern mathematics we wish to express the above procedure of treating symmetry in an axiomatic form; that is a form which does not depend on the particular example that we have used in our discussion. To this end we consider the essential ideas involved in our discussion of the symmetry of the equilateral triangle.

The symmetry appears to be described by 6 motions symbolised by a, a^2, b, ab, a^2b, and $a^3 = b^2 = e$; and the way in which these 6 motions can be combined in a kind of product. For example: a^2b followed by ab has the same effect (starting from the same initial position, such as that in Figure 1.1.1) as a^2.

We can represent this by writing $ab.a^2b = a^2$.

In order to express this in an abstract way we need:

(1) a set G (of motions)

(2) a product defined on G; that is a rule which assigns a unique element z of G to a given ordered pair of elements x, y of G. In other words, we want a binary operation on G.

This is often expressed by saying that there is a function $f\colon G \times G \to G$, where $f((x, y)) = z$. We shall write $f((x, y)) = xy$.

But this is not enough. We also need some axioms telling us how the product behaves. For example, if we were talking about a product of integers, then we would expect to have an axiom which said:

$$x(yz) = (xy)z$$

Because, if x, y and z are interpreted as integers and xy is the usual product of x and y, then the associative rule holds; in particular with $x = 3$, $y = 7$, $z = 2$, we have: $3 \times (7 \times 2) = (3 \times 7) \times 2$.

Our next task, therefore, is to find the rules which govern the behaviour of the product in our case. Now for us $x(yz)$ means the motion z followed by the motion y and then the result followed by the motion x. This is exactly the same as $(xy)z$. Thus, $(xy)z = x(yz)$. Hence we want the associative rule just as for the usual product of integers.

For integers, $xy = yx$, but we note that in our case $ab \neq ba$. Hence the commutative rule does not hold for us. However, in our way of using motions to express the symmetry of an object in space, it is clear that we must always have an identity motion, as represented by the symbol e in our example, with the property that $ex = x = xe$ for all x. Moreover for each motion x there must be an inverse motion which undoes the effect of x. For example, in the symmetry of the equilateral triangle $ba^2.ab = e$. Hence ba^2 undoes the effect of ab. If x denotes the motion, then we use the symbol x^{-1} to denote the motion which undoes the effect of x. We call x^{-1} the inverse of x. From $x^{-1}.x = e = x.x^{-1}$ we deduce that if x^{-1} is the inverse of x, then x is the inverse of x^{-1}.

Putting together the various parts of the above discussion we arrive at the following attempt at an abstract formulation of the intuitive notion of symmetry.

1.2 The Definition of a Group

A group consists of:

(1) a set G,

(2) a product on G, where xy denotes the product

of the elements x and y of G; together with the following axioms.

(3) The associative rule $x(yz) = (xy)z$ holds for all $x, y, z \in G$.

(4) There exists e in G with the property that $ex = x = xe$ for all x in G. (e is unique and is called the identity of G. The proof of uniqueness is in Exercise 1.1.)

(5) To each $x \in G$ there exists x' in G with the property that $x'x = e = xx'$. (x' is unique and is called the inverse of x. The proof of uniqueness is in Exercise 1.1. Once uniqueness is established the inverse of x can be given a special symbol, namely x^{-1}.)

If we have also:

(6) $xy = yx$ for all $x, y \in G$, we call G an abelian or commutative group.

If G has a finite number of elements it is called a finite group, otherwise it is an infinite group.

Note: (1) Unless otherwise stated, e will always be used to denote the identity of the group.

(2) Indices are defined as follows:

$x^0 = e$, $x^n = x\ x\ x\ \ldots\ x$ (n factors), $x^{-n} = (x^{-1})^n$, $n > 0$. By the uniqueness of the inverse, $(x^n)^{-1} = (x^{-1})^n$.

We leave the reader to check that the usual laws of indices hold.

The group concept attempts to make precise our intuitive notion of symmetry. In fact, the above axioms 3, 4, 5 assert more than is actually required to define a group. This question is taken up in Exercises 1.2 and 1.3.

1.3 The General Associative Law

This law asserts that we can insert brackets in any way that makes sense in the product of $n \geq 3$ elements of the group without affecting the value of the product. For example:

$$((ab)c)(d(gf)) = (a((bc)(dg)))f \quad .$$

PROOF The required result is true when $n = 3$. This is just the ordinary associative law.

Suppose the result is true for products of less than n elements. We consider a product of n elements. Suppose it has been calculated in one way to give a final product of 2 factors:

$$(a_1 a_2 \dots a_r)(a_{r+1} a_{s+2} \dots a_n)$$

and in another way to give a product of the 2 factors:

$$(a_1 a_2 \dots a_s)(a_{s+1} a_{s+2} \dots a_n) \quad .$$

Let $r \leqq s$. Notice that the products within each bracket involve less than n factors so are well-defined whatever the positions of the brackets within them. If $r = s$, the required result follows at once. If $r < s$, since $s < n$, we can write:

$$\begin{aligned}(a_1 a_2 \dots a_s)(a_{s+1} \dots a_n) &= ((a_1 a_2 \dots a_r)(a_{r+1} \dots a_s))(a_{s+1} \dots a_n)\\ &= (a_1 a_2 \dots a_r)((a_{r+1} \dots a_s)(a_{s+1} \dots a_n)),\end{aligned}$$

by ordinary associativity,

$$= (a_1 a_2 \dots a_r)(a_{r+1} \dots a_n) \quad .$$

Thus the required result holds for products of n elements if it holds for products of less than n elements. An appeal to induction completes the proof.

1.4 Further Examples of Groups

We derived the concept of a group from our consideration of the symmetry of the equilateral triangle. At the moment this provides us with our only example of a group. Let us pause for a moment to remedy this.

We leave the reader to check that the following are indeed examples of groups.

(1) The set of all non-singular $n \times n$ matrices over the real numbers under ordinary matrix multiplication. This is called the general linear group of degree n over $\mathbf{R}$ and is denoted by $GL_n(\mathbf{R})$ (see [G].)

(2) The set of integers under addition. This group is abelian and is denoted by $(\mathbf{Z}, +)$.

(3) The set of residue classes of integers modulo n under addition, denoted by $(\mathbf{Z}_n, +)$. This group is constructed as follows.

Take $\mathbf{Z}$ and define on it an equivalence relation $\equiv$ as follows. $a \equiv b$ if and only if n divides $a - b$, written $n \mid (a - b)$.

The equivalence class containing a we write as $\bar{a}$. We define addition of classes by $\bar{a} + \bar{b} = \overline{(a + b)}$.

The reader should check that this is a valid definition bearing in mind that $\bar{a} = \bar{a}'$, whenever $n \mid (a - a')$, and that our definition has been given in terms of particular elements of the equivalence classes involved.

(4) The set $\{\bar{1}, \bar{2}, \bar{3}, \ldots, \overline{(p - 1)}\}$ of non-zero elements of $\mathbf{Z}_p$, the set of residue classes of integers modulo a prime p, under the multiplication: $\bar{a}.\bar{b} = \overline{(ab)}$.

This group is denoted by $\mathbf{Z}_p^* = (\mathbf{Z}_p - \{\bar{0}\}, .)$. Here $\mathbf{Z}_p - \{\bar{0}\}$ means the set $\mathbf{Z}_p$ less the zero $\bar{0}$. If $p = 5$, we have:

$$\mathbf{Z}_p^* = (\mathbf{Z}_5 - \{\bar{0}\}, .) = \{\bar{1}, \bar{2}, \bar{3}, \bar{4}\}$$

(5) The set $\{1, i, -1, -i\}$ under the usual product of complex numbers. Note that all members of this group may be written in terms of i as follows: $\{i, i^2, i^3, i^4\}$. Such a group is said to be cyclic. All its members are powers of a single member.

(6) The set of 2 × 2 matrices:

$$\left\{ \begin{pmatrix} 1 & 0 \\ 0 & 1 \end{pmatrix}, \begin{pmatrix} 0 & 1 \\ -1 & 0 \end{pmatrix}, \begin{pmatrix} -1 & 0 \\ 0 & -1 \end{pmatrix}, \begin{pmatrix} 0 & -1 \\ 1 & 0 \end{pmatrix} \right\}$$

under the usual multiplication of matrices.

(7) If F is a field then $F^* = F - \{0\}$ is a group under the multiplication in the field F (see (4) above).

(8) Let F be any field. Then F is a group under the addition in the field. We denote this group by $(F, +)$.

1.5 Aims

In any science one of the major preoccupations is the classification of the objects of study. Another preoccupation is the investigation of structure; that is the way in which the objects of interest are constructed from simpler objects and how this affects their properties.

In group theory the object of study is the group. Hence the major aims of group theory are to classify the different types of group and to see how groups can be constructed from other groups which are simpler according to some well-defined criteria.

However, in a short introductory book like this, we can treat only a few elementary cases. Some of these will be discussed in the text, others will occur in exercises. In the course of following these aims many ideas, methods, and results will be considered which will prove useful in the applications of group theory to other branches of mathematics and to other sciences.

Let us return now to a closer study of the symmetry group of the equilateral triangle. This study will provide us with some basic ideas and techniques which will enable us to answer immediately the problem of classification for groups which have 1, 2, 3, or 4 elements.

EXERCISES 1

1. Let G be a group as defined in section 1.2. Prove that the identity e is unique and that the inverse of $a \in G$ is unique.

2. Let G be a set together with a binary operation $f\colon\ G \times G \to G$, where $f((a, b))$ is written ab. Let the following axioms hold:

(i) $(ab)c = a(bc)$ for all $a, b, c \in G$.

(ii) There exists a **left** identity $e \in G$ such that $ea = a$ for all $a \in G$.

(iii) For each $a \in G$ there exists a **left** inverse b such that $ba = e$.

Prove that G is a group with multiplication defined by f.

3. Let G be a set with binary operation as in question 2 above. Let the following axioms hold:

(i) $(ab)c = a(bc)$ for all $a, b, c \in G$.

(ii) The equations $ax = b$ and $ya = b$ can be solved for $x, y \in G$, where a and b are arbitrary members of G.

Prove that G is a group with multiplication defined by f. Note that a group as defined in section 1.2 satisfies the axioms given above.

4. Give an example of each of the following:

(i) an infinite non-abelian group,
(ii) an infinite abelian group,
(iii) a finite non-abelian group,
(iv) a finite abelian group.

CHAPTER 2.
MULTIPLICATION TABLE, GENERATORS, RELATIONS, ISOMORPHISM

2.1 <u>Multiplication Table</u>

In the group of symmetries of the equilateral triangle considered in section 1.1, we have 6 possible positions of symmetry obtained from the standard starting position of Figure 1.1.1 by the 6 motions symbolised by e, a, a^2, b, ab, a^2b.

In order to gain insight into the structure of the group we study the way in which these motions (or positions) combine under the product which was defined in section 1.1.1. To do this, it is convenient to use a device due to Cayley called the multiplication table of the group.

First we describe this in general terms, then we consider the particular case of the group mentioned above.

The elements of the group are written along the top row and down the left column of the table as references. The results of the multiplications are placed in the body of the table. We use $x_1, x_2, x_3, x_4, \ldots, x_n, \ldots$ as elements of the group. x_3x_2 is the product of x_3 and x_2. Of course, x_3x_2 is one of the elements $x_1, x_2, x_3, \ldots$ and is entered as such when the actual value is known.

<u>Note</u>: It is customary to take x_1 as the identity of the group. Usually the same order $x_1, x_2, x_3, \ldots$ is used to label the rows and the columns of the table.

Column

	x_1	x_2	x_3	x_4	....
x_1	x_1^2	x_1x_2	x_1x_3	x_1x_4	
x_2	x_2x_1	x_2^2	x_2x_3	x_2x_4	
Row x_3	x_3x_1	x_3x_2	x_3^2	x_3x_4	
x_4	x_4x_1	x_4x_2	x_4x_3	x_4^2	
....					

Table 2.1.1

2.2 **Multiplication Table for Dihedral Group D_3**

The symmetry group of the equilateral triangle is usually called the dihedral group D_3. In this notation D_n is the dihedral group with $2n$ elements. It is the symmetry group of a plane regular polygon with n sides. But more of this later. Let us consider D_3.

As we know, its elements can be written: e, a, a^2, b, ab, a^2b. We could go back to our diagrams or to a cardboard cut-out triangle in order to work out the various products. However, there is a purely algebraic method which uses the fact that: $a^3 = e$, $b^2 = e$, $ba = a^2b$. For example to calculate $ab.a^2b$ we have the sequence: $ab.a^2b = ab.ba = ab^2a = aea = a^2$, using the above equations.

Proceeding in this way we arrive at the following Table 2.2.1. We say that the group D_3 is generated by a and b subject to the relations $a^3 = e$, $b^2 = e$, $ba = a^2b$. We write:

$$D_3 = \langle a, b \mid a^3 = e, b^2 = e, ba = a^2b \rangle .$$

	e	a	a^2	b	ab	a^2b
e	e	a	a^2	b	ab	a^2b
a	a	a^2	e	ab	a^2b	b
a^2	a^2	e	a	a^2b	b	ab
b	b	a^2b	ab	e	a^2	a
ab	ab	b	a^2b	a	e	a^2
a^2b	a^2b	ab	b	a^2	a	e

Table 2.2.1

From this table we can read off the inverse of each element. If we want the inverse of the element x, where x is one of e, a, a^2, b, ab, a^2b, we run along the x-row until we encounter e. Then we run up the column to identify the column as, say, the y-column. Then the inverse of x is y.

Using this method we obtain the following table of inverses.

Element	Inverse
e	e
a	a^2
a^2	a
b	b
ab	ab
a^2b	a^2b

Table 2.2.2

Note: We notice another fact about the multiplication table. Each element of the group appears once and only once in each row of the table.

A similar remark is true for columns.

The proofs of these statements are given as exercises. (For several more interesting results on multiplication tables see: "Groups and their Graphs" - Grossman and Magnus (Random House 1964).)

2.3 Order of an Element

If we take some element of the group D_3, say, a^2, we note that:

$$(a^2)^2 = a,$$

$$(a^2)^3 = e.$$

In general if x is any element of the group D_3 we have $x^r = e$ for some positive integer r.

In fact this is true in any finite group. The proof of this is given as an exercise.

The smallest positive r for which $x^r = e$ is called the **order** of the element x. If no such r exists x is said to have infinite order. We shall have more to say about this later.

The order of an element must be distinguished from the order of the group. The latter is just the number of elements in the group, $|G|$. The order of D_3 is 6.

From the multiplication table we can quickly obtain the orders of the elements as given below.

Element	Order
e	1
a	3
a^2	3
b	2
ab	2
a^2b	2

Table 2.3.1

Comparing this table with Table 2.2.2, we notice that an element and its inverse have the same order. The proof of this in the general case is given as an exercise.

Remembering that the order of D_3 is 6, we observe that the orders of the elements are all divisors of 6. Could it be that there is a theorem here to the effect that for a finite group the order of an element divides the order of the group?

This question will be taken up later. This is the kind of relation that we must look for. Apart from its intrinsic interest it may help us in our main aims of classification and structure.

2.4 **The Symmetric Group S_n**

In order to introduce the important idea of isomorphism, and for use later, we are going to consider another example of a group.

Take the set $S = \{1, 2, 3, 4, 5, \ldots, n\}$. A permutation is a function $f;\ S \to S$, which is one-one and onto. Suppose:

$$f(1) = a_1,\quad f(2) = a_2,\quad f(3) = a_3,\ \ldots,\ f(n) = a_n,$$

where $a_1, a_2, a_3, \ldots, a_n$ are $1, 2, 3, 4, \ldots, n$ in some order. It is customary to represent f by:

$$\begin{pmatrix} 1 & 2 & 3 & 4 & \cdots & n \\ a_1 & a_2 & a_3 & a_4 & \cdots & a_n \end{pmatrix}$$

where the image $f(i)$ is positioned immediately below i.

In fact the order of the numbers in the top row does not matter as long as the right image $f(i)$ is below i. Thus we could write:

$$\begin{pmatrix} b_1 & b_2 & b_3 & \cdots & b_n \\ f(b_1) & f(b_2) & f(b_3) & \cdots & f(b_n) \end{pmatrix}$$

where $b_1, b_2, \ldots, b_n$ are $1, 2, 3, 4, \ldots, n$ in some order.

For example:

$$\begin{pmatrix} 1 & 2 & 3 \\ 2 & 3 & 1 \end{pmatrix},$$

where $n = 3$, is the function

$$f\colon\ \begin{array}{l} 1 \to 2 \\ 2 \to 3 \\ 3 \to 1 \end{array}$$

that is:

$$f(1) = 2$$
$$f(2) = 3$$
$$f(3) = 1 \quad .$$

This could also be expressed as:

$$\begin{pmatrix} 2 & 1 & 3 \\ 3 & 2 & 1 \end{pmatrix} \quad .$$

This permutation is an example of a special kind of permutation called a **cycle**. In general a cycle is a permutation of the form:

$$\begin{pmatrix} b_1 & b_2 & b_3 & \cdots & b_r \\ b_2 & b_3 & b_4 & \cdots & b_1 \end{pmatrix} ,$$

where $r \leq n$. The number r is called the **length** of the cycle.

This is usually written more briefly as:

$$(b_1 \; b_2 \; b_3 \; \cdots \; b_r) \quad .$$

In the example above we write $\begin{pmatrix} 1 & 2 & 3 \\ 2 & 3 & 1 \end{pmatrix}$ as $(1 \; 2 \; 3)$.

The length of this cycle is 3.

Another example:

$$\begin{pmatrix} 1 & 5 & 9 & 6 \\ 5 & 9 & 6 & 1 \end{pmatrix} = (1 \quad 5 \quad 9 \quad 6)$$

is a cycle on $S = \{1, 2, 3, 4, 5, 6, 7, 8, 9, 10\}$ of length 4.

Note that cycles of length 1 mean that the symbols involved are left fixed. Thus (m) denotes that the symbol m is fixed in the permutation. Such cycles are often omitted when describing the permutation.

There are $n!$ distinct permutations for a set S with n members. We denote the set of all $n!$ permutations on S by S_n.

A product may be defined on S_n as follows. Let f and g be two permutations $S \to S$, that is $f \in S_n$ and $g \in S_n$. The product gf of the two functions f and g is just the composition of the functions, namely first carry out the function f, then follow this by carrying out the function g. Thus gf: $S \overset{f}{\to} S \overset{g}{\to} S$ is given by

$$(gf)(i) = g(f(i)), \qquad (i = 1, 2, 3, 4, \ldots, n).$$

Note: some authors reverse the order here, and by gf they mean first g then f. We follow a notation which is consistent with the functional notation used in the calculus for a function of a function with the function symbol on the left of the variable. Although there are advantages in group theory in the alternative notation, our notation seems to suit most people who are not intending to specialise in group theory.

For example, if $S = \{1, 2, 3\}$ and f is the permutation:

$$\begin{pmatrix} 1 & 2 & 3 \\ 2 & 1 & 3 \end{pmatrix} = (12)(3) = (12),$$

while g is: (13). Then

$$\begin{aligned} (gf)(1) &= g(f(1)) = g(2) = 2 \\ (gf)(2) &= g(f(2)) = g(1) = 3 \\ (gf)(3) &= g(f(3)) = g(3) = 1\,. \end{aligned}$$

Thus:

$$(gf) = \begin{pmatrix} 1 & 2 & 3 \\ 2 & 3 & 1 \end{pmatrix} = (1\ 2\ 3)\,.$$

The following notation is suggestive:

$$(gf) = \begin{pmatrix} 2 & 1 & 3 \\ 2 & 3 & 1 \end{pmatrix}\begin{pmatrix} 1 & 2 & 3 \\ 2 & 1 & 3 \end{pmatrix} = \begin{pmatrix} 1 & 2 & 3 \\ 2 & 3 & 1 \end{pmatrix}\,.$$

We prove now that S_n, with the product defined as above, is a group of order $n!$.

First the definition given above is a valid definition of a binary operation on S_n; that is, it is a function from

$$S_n \times S_n \to S_n$$

$$(g, f) \to gf$$

To each ordered pair of permutations (g, f) we have defined a unique permutation gf.

Next there is an identity element in S_n. The obvious way of seeing this is to nominate a likely candidate and then verify that this candidate satisfies the defining conditions for an identity. We propose $\begin{pmatrix} 1 & 2 & 3 & \cdots & n \\ 1 & 2 & 3 & \cdots & n \end{pmatrix}$ as the identity. We have:

$$\begin{pmatrix} 1 & 2 & 3 & \cdots & n \\ a_1 & a_2 & a_3 & & a_n \end{pmatrix}\begin{pmatrix} 1 & 2 & 3 & \cdots & n \\ 1 & 2 & 3 & \cdots & n \end{pmatrix} = \begin{pmatrix} 1 & 2 & 3 & \cdots & n \\ a_1 & a_2 & a_3 & \cdots & a_n \end{pmatrix}$$

and

$$\begin{pmatrix} 1 & 2 & 3 & \cdots & n \\ 1 & 2 & 3 & \cdots & n \end{pmatrix}\begin{pmatrix} 1 & 2 & 3 & \cdots & n \\ a_1 & a_2 & a_3 & \cdots & a_n \end{pmatrix} = \begin{pmatrix} a_1 & a_2 & \cdots & a_n \\ a_1 & a_2 & \cdots & a_n \end{pmatrix}\begin{pmatrix} 1 & 2 & \cdots & n \\ a_1 & a_2 & \cdots & a_n \end{pmatrix}$$

$$= \begin{pmatrix} 1 & 2 & 3 & \cdots & n \\ a_1 & a_2 & a_3 & \cdots & a_n \end{pmatrix} .$$

Thus if $f = \begin{pmatrix} 1 & 2 & 3 & \cdots & n \\ a_1 & a_2 & a_3 & \cdots & a_n \end{pmatrix}$ is an arbitrary permutation in S_n and $e = \begin{pmatrix} 1 & 2 & 3 & \cdots & n \\ 1 & 2 & 3 & \cdots & n \end{pmatrix}$, we have:

$$ef = f = fe .$$

These are the defining equations for e to be an identity of S_n.

It is usual to denote $\begin{pmatrix} 1 & 2 & 3 & \cdots & n \\ 1 & 2 & 3 & \cdots & n \end{pmatrix}$ by (1).

Finally, given any permutation $f \in S_n$, we have to find a permutation g with the property that

$$gf = e = fg .$$

Let $f = \begin{pmatrix} 1 & 2 & 3 & \cdots & n \\ a_1 & a_2 & a_3 & \cdots & a_n \end{pmatrix}$, we propose $g = \begin{pmatrix} a_1 & a_2 & \cdots & a_n \\ 1 & 2 & \cdots & n \end{pmatrix}$.

We have:

$$gf = \begin{pmatrix} a_1 & a_2 & \cdots & a_n \\ 1 & 2 & \cdots & n \end{pmatrix} \begin{pmatrix} 1 & 2 & \cdots & n \\ a_1 & a_2 & \cdots & a_n \end{pmatrix} = \begin{pmatrix} 1 & 2 & 3 & \cdots & n \\ 1 & 2 & 3 & \cdots & n \end{pmatrix} = e ,$$

and

$$fg = \begin{pmatrix} 1 & 2 & \cdots & n \\ a_1 & a_2 & \cdots & a_n \end{pmatrix} \begin{pmatrix} a_1 & a_2 & \cdots & a_n \\ 1 & 2 & \cdots & n \end{pmatrix} = \begin{pmatrix} a_1 & a_2 & \cdots & a_n \\ a_1 & a_2 & \cdots & a_n \end{pmatrix}$$

$$= \begin{pmatrix} 1 & 2 & \cdots & n \\ 1 & 2 & \cdots & n \end{pmatrix} = e .$$

Thus g is an inverse of f. But we already know that such an inverse is unique, hence g is the inverse of f and we may write

$$g = f^{-1} = \begin{pmatrix} a_1 & a_2 & \cdots & a_n \\ 1 & 2 & \cdots & n \end{pmatrix} .$$

The defining axioms for a group are all satisfied. Hence S_n with the given product is a group of order $n!$. In general S_n is not an abelian group. This can be seen as follows.

Suppose $n \geqq 3$. Take (13) and (123) $\in S_n$. Then (13)(123) = (12), however, (123)(13) = (23). Thus (13)(123) $\neq$ (123)(13). Hence S_n, $n \geqq 3$, is not abelian.

S_n is called the symmetric group on n symbols. By suitable labelling, the symbols can always be taken to be 1, 2, 3, 4, 5, ..., n.

Let us consider now a special case of S_n, namely, S_3. This group has order 3! = 6. It turns out to be the smallest non-abelian group, as we shall see later. As such it is very useful as a simple example of the various concepts which we shall meet.

We may write:

$$S_3 = \{(1), (12), (13), (23), (123), (132)\} .$$

This is just the set of all permutations on the set of 3 symbols, say {1, 2, 3} .

Working out products such as (123)(23) according to the definition, we have f = (23) followed by g = (123). Thus:

$$(gf)(1) = g(f(1)) = g(1) = 2 ,$$
$$(gf)(2) = g(f(2)) = g(3) = 1 ,$$
$$(gf)(3) = g(f(3)) = g(2) = 3 .$$

Thus:

$$(123)(23) = gf = \begin{pmatrix} 1 & 2 & 3 \\ 2 & 1 & 3 \end{pmatrix} = (12) .$$

In this way we construct the multiplication table for S_3.

	(1)	(123)	(132)	(12)	(13)	(23)
(1)	(1)	(123)	(132)	(12)	(13)	(23)
(123)	(123)	(132)	(1)	(13)	(23)	(12)
(132)	(132)	(1)	(123)	(23)	(12)	(13)
(12)	(12)	(23)	(13)	(1)	(132)	(123)
(13)	(13)	(12)	(23)	(123)	(1)	(132)
(23)	(23)	(13)	(12)	(132)	(123)	(1)

Multiplication Table 2.4.1. for S_3

2.5 **Isomorphism**

If we write a for (123), b for (12) and e for (1), the above table becomes identical with Table 2.2.1, the multiplication table of D_3, the group of symmetries of the equilateral triangle obtained in section 2.2.

We describe this situation by saying that D_3 is **isomorphic** to S_3. In general two groups are said to be isomorphic if the elements correspond one to one so that the two multiplication tables are identical. Note that this depends on the corresponding elements being written in similar positions in the two tables. This can be expressed more precisely, and in a form suitable for infinite groups, as follows.

DEFINITION 2.5.1. *Two groups G and H are isomorphic (symbolised by $G \cong H$) if and only if there exists a function $f: G \to H$ such that*

(1) f is one-one, that is: $f(g_1) = f(g_2) \to g_1 = g_2$; $g_1, g_2 \in G$

(2) f is onto, that is: for each $h \in H$, there exists $g \in G$, such that $f(g) = h$.

(3) $f(g_1g_2) = f(g_1)f(g_2)$, *for all* $g_1, g_2 \in G$, *where* g_1g_2 *is the product of* g_1 *and* g_2 *in* G *and* $f(g_1)f(g_2)$ *is the product of* $f(g_1)$ *and* $f(g_2)$ *as elements of* H.

Conditions (1) and (2) above give the one-one correspondence required between the elements of G and H, and (3) ensures that the multiplication tables can be identified using this correspondence of elements. Condition (3) is often expressed by saying that the function f preserves the structure, in this case of a group.

Note that as far as we are concerned in this book isomorphic groups are to be identified. This is important when we consider a question such as: how many groups of order 28, say, are there?

In fact one of the main difficulties when answering this question for groups of moderate, but not very small order, is to ensure that the same group is not turning up in more than one form.

2.6 Generators and Relations

Consider the multiplication table we obtain according to the following construction.

We are given two symbols a and b with which we form words or strings of symbols such as:

$$aba^2b^2a^3b^5a^4$$

where a^5 is short for $aaaaa$. We write e for the string with no symbols in it, the so-called empty string. We are now given rules, or relations, which enable us to replace one string by another which is to be considered equal to it. For example, the relation $a^2 = e$ would mean that a^2 could be replaced by e to give an equal word. Thus $aba^2b^2a^3b^5a^4$ would be considered the same as $abeb^2aeb^5e^2$. Moreover, since e is the empty word or string, this is the same as ab^3ab^5.

Let us consider the multiplication table we obtain according to the description:

$$\langle a, b \mid a^3 = e, \quad b^2 = e, \quad ba = a^2b \rangle$$

Here a, b are the **generators** and $a^3 = e$, $b^2 = e$, $ba = a^2b$ are the **defining relations**.

Note the notation $\langle \ldots \mid \ldots \rangle$. The generators are written on the left of the line | and the relations on the right. Altogether the expression

$$\langle a, b, c, \ldots \mid R_1, R_2, \ldots \rangle$$

denotes the group **generated** by the symbols a, b, ... subject to the relations R_1, R_2, ... according to the procedure described below. In the general case some strings will involve symbols a^{-1}, b^{-1}, ... etc. and their powers. We have avoided this by using relations of the form $a^r = e$, where $r > 0$, $a^{-1} = a^{r-1}$.

It must be emphasised that the treatment here is **informal**. Some idea of the **rigorous** treatment may be obtained from sections 5.1 to 5.4, where the theory is developed for **abelian** groups.

First we notice that every string may be reduced to one of the following:

$$e, a, a^2, b, ab, a^2b$$

by using the defining relations. Moreover no two of these are equal. For example, if $ab = a^2b$ then $abb = a^2bb$. Now this gives $ae = a^2e$. Thus $a = a^2$. But then $a^2a = a^2a^2$, which reduces to $e = a^3a$. Thus $e = ea = a$, a contradiction, because a is not the empty string.

Using the relations $a^3 = e$, $b^2 = e$, $ba = a^2b$ we can calculate products of the 6 distinct strings. For example $ab.a^2b = aa^2bab = a^3a^2bb = ea^2e = a^2$. We are, of course, assuming the associative rule throughout.

In this way we obtain a multiplication table identical with 2.2.1. This means that we have constructed a group, isomorphic to D_3 and S_3, in which the elements are strings of symbols.

We call this a **finite presentation** (or a **generators** and **relations description**) of D_3 and S_3.

It can be proved that this kind of procedure, defined more precisely of course, always leads to a group. Every relation involving the generators can be obtained by algebraic manipulation from the defining relations.

For a lucid introduction to the **rigorous** treatment, which involves the theory of free groups, see: "Presentation of Groups" - Johnson (C.U.P. 1976).

This branch of group theory is often called '**combinatorial group theory**'. Other references are given in the bibliography at the end of the book.

2.7 All Possible Groups of Orders 1, 2, 3, or 4

We are now in a position to find all possible groups of orders 1, 2, 3, or 4. In finding such groups we make the point once again that we do not distinguish between isomorphic groups.

We note two facts about the multiplication table of a group. (The first has been mentioned already.) For convenience of future reference we exalt these to the rank of:

THEOREM 2.7.1. *Let G be a group.*

(1) Each element of the group appears once and only once in each row and column of the table.

(2) If $ab = e$, then $ba = e$; $a, b \in G$.

PROOF The proof of (1) is given as exercise 2.2.

If $ab = e$, then $abb^{-1} = b^{-1}$. Hence $a = b^{-1}$.
Thus

$$ba = bb^{-1} .$$

Finally $ba = e$.

Order 1. Upto isomorphism there is only one group of order 1, namely $G = \{e\}$, with multiplication table:

	e
e	e

Table 2.7.1

Order 2. By theorem 2.7.1 the multiplication table must be:

	e	a
e	e	a
a	a	e

Table 2.7.2

This is a special case of a cyclic group; that is a group whose elements are all powers of a single element. Such a group of order n will consist of the elements $\{e, a, a^2, a^3, \ldots, a^{n-1}\}$, where $ea = a = ae$, and with multiplication $a^r.a^s = a^{(r+s)}$, where $(r + s)$ means that $r + s$ is reduced modulo n.

For example, if $n = 6$, $r = 3$, $s = 5$, then $r + s = 8$, which reduces modulo 6 to 2. Thus $(r + s) = 2$.

In general we write: $r + s = qn + t$, where $0 \leq t < n$, so that $(r + s) = t$. If $t = 0$, we put $a^t = a^0 = e$.

The multiplication just defined is associative as can be seen from the equations:

$$a^r(a^s a^t) = a^r(a^{(s+t)}) = a^{(r+(s+t))} = a^{(r+s+t)} \quad \text{and}$$

$$(a^r a^s)a^t = a^{((r+s)+t)} = a^{(r+s+t)} \, .$$

The inverse of a^r is a^{n-r}.

Such a finite group can be described in terms of generators and relations by:

$$\langle a \mid a^n = e \rangle \quad .$$

We denote this group by C_n .

An infinite cyclic group consists of the elements

$$\{e,\ a^{\pm r} \mid r = 1,\ 2,\ 3,\ \ldots\} = \langle a \mid \ \rangle ,$$

in terms of the generator a. We have the following theorem:

THEOREM 2.7.2. *All cyclic groups of order n are isomorphic to* $(\underline{\mathbf{Z}}_n,\ +)$. *All infinite cyclic groups are isomorphic to* $(\underline{\mathbf{Z}},\ +)$. (See section 1.4 examples (2) and (3) for definitions of $(\underline{\mathbf{Z}},\ +)$ and $(\underline{\mathbf{Z}}_n,\ +)$.)

PROOF Let $C_n = \{e,\ a,\ a^2,\ \ldots,\ a^{n-1}\} = \langle a \mid a^n = e \rangle$. Define a function

$$f\colon\ C_n \to (\underline{\mathbf{Z}}_n,\ +) = \{\bar{0},\ \bar{1},\ \bar{2},\ \ldots,\ \overline{(n-1)}\}$$

by: $f(a^r) = \bar{r}$. Thus:

$$f\colon\quad \begin{array}{lll} e \to \bar{0} & a^3 \to \bar{3} & a^6 \to \bar{6} \\ a \to \bar{1} & a^4 \to \bar{4} & \vdots \\ a^2 \to \bar{2} & a^5 \to \bar{5} & a^r \to \bar{r} \end{array} \quad \text{and so on .}$$

Hence f is one-one and onto.

Let $0 \leqq r \leqq n-1$ and $0 \leqq s \leqq n-1$. Then:

$$\begin{aligned} f(a^r.a^s) &= f(a^{r+s}) = f(a^t)\ , && \text{where } 0 \leqq t \leqq n-1 \\ & && \text{and } r + s \equiv t \quad (\text{modulo } n), \\ &= \bar{t} = \bar{r} + \bar{s}\ , && \text{because } r + s \equiv t \quad (\text{modulo } n), \\ &= f(a^r) + f(a^s)\ . \end{aligned}$$

Note + is used to denote the product in $(\underline{\mathbf{Z}}_n,\ +)$. Thus f is an isomorphism of C_n with $(\underline{\mathbf{Z}}_n,\ +)$.

Now let C_∞ be an infinite cyclic group:

$$\{e,\ a^{\pm 1},\ a^{\pm 2},\ a^{\pm 3},\ \ldots,\ a^{\pm r},\ \ldots\} = \langle a \mid \rangle .$$

Define a function

$$f\colon\ C_\infty \to (\mathbf{Z}, +) = \{0, \pm 1, \pm 2, \pm 3, \ldots\}$$

by $\quad f(a^r) = r$. Thus:

$$f\colon\ \begin{matrix} e \to 0 & a^2 \to 2 & a^{-3} \to -3 \\ a \to 1 & a^{-2} \to -2 & a^4 \to 4 \\ a^{-1} \to -1 & a^3 \to 3 & a^{-4} \to -4 \end{matrix} \quad \text{and so on,}$$

is one-one and onto. Moreover,

$$f(a^r.a^s) = f(a^{r+s}) = r + s = f(a^r) + f(a^s) .$$

Thus C_∞ is isomorphic to $(\mathbf{Z}, +)$.

Note that this completely classifies all cyclic groups. We shall attend to their structure later.

Order 3. Straightforward application of theorem 2.7.1 gives just one possible multiplication table describing a cyclic group:

$$\langle a \mid a^3 = e \rangle = \{e, a, a^2\}$$

Order 4. Application of theorem 2.7.1 leads this time to 4 possible multiplication tables according to the position of the identity e in the second and third rows of the tables. However, there are only 2 non-isomorphic groups. One is the cyclic group:

$$\langle a \mid a^4 = e \rangle = \{e, a, a^2, a^3\} .$$

The other is the Klein 4-group

$$K_4 = \langle a, b \mid a^2 = b^2 = e,\ ab = ba \rangle = \{e, a, b, ab = ba\} ,$$

with multiplication table 2.7.3.

	e	a	b	ab
e	e	a	b	ab
a	a	e	ab	b
b	b	ab	e	a
ab	ab	b	a	e

Table 2.7.3

K_4 is isomorphic to the group of permutations with multiplication table 2.7.4. This latter group is contained in S_4 and has the same rule for forming products as S_4, namely the usual product of permutations. It is an example of a **subgroup** of S_4. We return to this idea later.

	(1)	(12)	(34)	(12)(34)
(1)	(1)	(12)	(34)	(12)(34)
(12)	(12)	(1)	(12)(34)	(34)
(34)	(34)	(34)(12)	(1)	(12)
(12)(34)	(12)(34)	(34)	(12)	(1)

(12)(34) = (34)(12)

Table 2.7.4

We leave the details of the above to the reader. These results can be obtained much more quickly once Lagrange's theorem becomes available in Chapter 3.

To sum up, we have shown that, upto isomorphism, there are just 2 groups of order 4; the cyclic group C_4 of order 4 and the Klein 4-group K_4.

2.8 Some Results on Orders of Elements

As an epilogue to this chapter we collect together some results on the orders of elements in a finite group. They will prove of considerable value to us in our future work. Some of these have intrinsic interest. For ease of reference we call this collection of results:

THEOREM 2.8.1. *Let G be a group. Denote the order of the element $g \in G$ by $O(g)$. Denote the greatest common divisor of n and m by (n, m).*

(1) If $g \in G$ has order n, then g^{-1} has order n.

(2) If $g \in G$ has order n, then $g^m = e$ if and only if n divides m, denoted by $n \mid m$.

(3) If $b = x^{-1}ax$, for $x, a, b \in G$, then a, and b have the same order.

(4) Let $O(a) = n$, $O(a^r) = m$, and $(n, r) = d$, then $m = n/d$.

(5) Let $O(a) = n$, $O(b) = m$, $(m, n) = 1$, and $ab = ba$, then $O(ab) = mn$.

(6) Let $O(g) = mn$, and $(m, n) = 1$, then $g = ab = ba$, for some $a, b \in G$, where $O(a) = n$, $O(b) = m$. Such an expression for g is unique.

(7) If $(a_1\ a_2\ a_3\ \dots\ a_r)$ is a cycle of length r in the symmetric group S_n, then $O((a_1\ a_2\ \dots\ a_r)) = r$.

(8) Let $f \in S_n$ be written as a product of disjoint cycles: $f = c_1\ c_2\ \dots\ c_t$. Then:

$O(f)$ = the lowest common multiple of:

$$O(c_1),\ O(c_2),\ O(c_3),\ \ldots,\ O(c_t)\ .$$

PROOF

(1) Exercise 2.4.

(2) Let $g^m = e$. Write $m = qn + r$, for integers q and r, with $0 \leq r < n$. Then:

$$e = g^m = g^{qn+r} = (g^n)^q . g^r = e . g^r = g^r\ .$$

Now $O(g) = n$, and $r < n$. Thus $r = 0$. Hence $n \mid m$.

Conversely, if $n \mid m$, we have:

$$m = qn\ .$$

Hence $g^m = g^{qn} = (g^n)^q = e^q = e\ .$

(3) Let $O(a) = n$ and $O(b) = m$. Then

$$b^n = (x^{-1}ax)^n = x^{-1}a^n x = x^{-1}ex = e\ .$$

Thus $m \leq n\ .$

Similarly $a^m = (xbx^{-1})^m = xb^m x^{-1} = e\ .$

Thus $n \leq m\ .$

Hence $n = m\ .$

(4) $(n\ ,\ r) = d$. Let $n = sd$ and $r = \ell d$, where $(s,\ \ell) = 1$. Then

$$(a^r)^s = a^{s\ell d} = (a^n)^\ell = e^\ell = e\ .$$

Now $m = O(a^r)$. Thus $m \mid s$. Also $e = (a^r)^m = a^{\ell dm}$. Hence

$$n \mid \ell dm\ .$$

Since $n = sd$, we conclude that $s \mid \ell m$.

Now $(s,\ \ell) = 1$. Thus $s \mid m$.

From $s \mid m$, $m \mid s$ and m, s both positive integers, we deduce that

$$m = s = n/d\ .$$

(5) $(ab)^{mn} = a^{mn}b^{mn} = e$, because $ab = ba$. Thus

$$O(ab) \mid mn\ .$$

Let $O(ab) = m_1 n_1$, where $m_1 \mid m$ and $n_1 \mid n$. Then

$$(m_1,\ n_1) = 1\ ,$$

and we may write

$$m = tm_1 ,$$

$$n = sn_1 ,$$

for some positive integers t and s.

Thus

$$e = (ab)^{tm_1n_1} = (ab)^{mn_1} = a^{mn_1}b^{mn_1} = a^{mn_1}.e = a^{mn_1} .$$

Hence $a^{mn_1} = e$. Therefore, by (2) above, $n \mid mn_1$.

Now $(n, m) = 1$. Hence $n \mid n_1$.

Since $n_1 \mid n$, and n and n_1 are positive integers, we conclude that $n = n_1$.

A similar argument shows that $m = m_1$. Thus

$$O(ab) = mn.$$

(6) Since $(m, n) = 1$, we have $Mm + Nn = 1$ for suitable integers M and N (see [G]). Hence

$$g = g^{Mm+Nn} = g^{Mm}.g^{Nn} .$$

Put $a = g^{Mm}$ and $b = g^{Nn}$. Then $g = ab = ba$.

Now since g has order mn, g^m must have order n. Thus $(g^m)^M$ has order n by (4) above, since from $Mm + Nn = 1$, we have $(M, n) = 1$. By a similar argument, g^{Nn} has order m. Thus

$$O(a) = n$$

and

$$O(b) = m$$

as required.

To prove uniqueness, suppose we also had $g = a_1b_1 = b_1a_1$ with $O(a_1) = n$ and $O(b_1) = m$. Then

$$a_1b_1 = ab .$$

Hence

$$(a_1b_1)^{Mm} = (ab)^{Mm} .$$

But $a_1b_1 = b_1a_1$ and $ab = ba$. Thus

$$a_1^{Mm}b_1^{Mm} = a^{Mm}b^{Mm} .$$

Now $O(b_1) = O(b) = m$. Hence

$$b_1^{Mm} = b^{Mm} = e .$$

Thus the above becomes:

$$a_1^{Mm} = a^{Mm} .$$

Using $Mm + Nn = 1$, we have $a_1^{1-Nn} = a^{1-Nn}$. Thus

$$a_1 . a_1^{-Nn} = a . a^{-Nn} .$$

$$a_1 = a . a^{-Nn} . a_1^{Nn} .$$

Now $O(a_1) = n$; hence $a_1^{Nn} = e$. Thus

$$a_1 = a . a^{-Nn} e = a . a^{-Nn} .$$

Then $a_1 a^{Nn} = a . a^{-Nn} . a^{Nn} = a$. Now $a^{Nn} = e$. Thus

$$a_1 = a .$$

Finally from $a_1 b_1 = ab$, we deduce that $b_1 = b$.

Altogether we have $a_1 = a$ and $b_1 = b$. The required uniqueness follows.

(7) $(a_1 a_2 a_3 \ldots a_r)^t$ first becomes equal to the identity permutation when $t = r$. Thus

$$O((a_1 a_2 \ldots a_r)) = r .$$

(8) Let $f = c_1 c_2 c_3 \ldots c_t$. Let $O(f) = m$. Let the L.C.M. of $O(c_1)$, $O(c_2)$, ..., $O(c_t)$, be ℓ. Then

$$f^\ell = (c_1 c_2 \ldots c_t)^\ell = c_1^\ell c_2^\ell \ldots c_t^\ell = e$$

because disjoint cycles commute and

$$c_i^\ell = e$$

for $i = 1, 2, \ldots, t$.

Thus $m \mid \ell$. Also

$$e = f^m = c_1^m c_2^m \ldots c_t^m .$$

Now c_1^m, c_2^m, c_3^m, c_4^m, ..., c_t^m involve distinct symbols. Therefore for $c_1^m c_2^m \ldots c_t^m$ to be equal to e, each c_i^m must **separately** be e.

Thus $\quad O(c_i) \mid m$.

Hence $\quad \ell \mid m$.

From $\ell \mid m$ and $m \mid \ell$ we conclude that $\ell = m$.

Note.

In this chapter, in order to give the reader something concrete with which to work, we have perhaps rather over-emphasised the importance of multiplication tables. To correct this we point out that such tables play a relatively small part in group theory as a whole. Similar remarks also apply to the study of groups of very small order. However, the latter do provide a useful vehicle for introducing ideas and aims that are of wide application and significance.

Chapter 8 (Epilogue) goes some way towards putting into perspective our preoccupation with multiplication tables and small order groups.

EXERCISES 2

1. Find the multiplication table for D_4, the group of symmetries of a square.

2. Prove that in the multiplication table of a finite group each element of the group appears once and only once in each row and column of the table.

3. Let G be a finite group. Let $x \in G$. Prove that $x^r = e$ for some positive integer r.

4. Let G be a group. Let x be an element of G of order n. Prove that the order of x^{-1} is also n.

5. Find the orders and inverses of the elements of D_4 (see question 1 above).

6. Let G be a cyclic group. Prove that G must be abelian.

7. Give an example to show that not all abelian groups are cyclic.

8. Let $f \in S_n$ be an arbitrary permutation. Prove that f may be written as a product of disjoint cycles. Carry out the process for the particular cases

$$f = \begin{pmatrix} 1 & 2 & 3 & 4 & 5 & 6 & 7 \\ 3 & 4 & 1 & 6 & 7 & 2 & 5 \end{pmatrix} \in S_7 \quad \text{and} \quad g = \begin{pmatrix} 1 & 2 & 3 & 4 & 5 & 6 \\ 1 & 5 & 4 & 3 & 6 & 2 \end{pmatrix} \in S_6 .$$

[**Hint**: Start $(1\ f(1)\ f^2(1)\ \ldots\ )$.]

9. Prove that every permutation $f \in S_n$ may be written as a product of cycles of length 2, <u>not</u> necessarily disjoint.

[**Note**: A cycle of length 2 is usually called a **transposition**.
Hint: First write an arbitrary cycle in the required form, then use the result of question 8.]

10. With the help of theorem 2.8.1, part (8), find the orders of f and g in question 8 above.

CHAPTER 3.
SUBGROUPS, LAGRANGE'S THEOREM, CYCLIC GROUPS

3.1 Cosets and Lagrange's Theorem

Consider a group G of order p, where p is a prime integer. Suppose for the moment that our conjecture in section 2.3 that 'the order of an element of a finite group divides the order of the group' has been proved. There are p elements in G so there exists $g \in G$ with $g \neq e$. Because G is finite, g has finite order, say $r > 1$. Moreover by the above supposition r divides p. Now p is prime and $r > 1$. Hence $r = p$.

Now the set of elements $\{g, g^2, g^3, \ldots, g^{p-1}, g^p = e\}$ is the cyclic group described by $\langle g \mid g^p = e \rangle$ as discussed in section 2.7. It has order p. Now the order of G is p. It follows that $G = \langle g \rangle$.

This determines all groups whose order is a prime. They are all cyclic groups.

We have made a small start in the search for the groups of a given order. Since it follows from our conjecture, we have at least some incentive to look for a proof of the latter.

Now in any group G, if we take an element $g \neq e$ of order r, the set of elements $\{g, g^2, \ldots, g^{r-1}, g^r = e\}$ is the cyclic group of order r described by $\langle g \mid g^r = e \rangle$. Any cyclic subgroup of G has this form for some $g \in G$. Thus if the order r of an element g of a finite group G divides the order of the group G, then the order of a cyclic subgroup of G divides the order of the group G; and conversely.

Thus our conjecture is equivalent to a conjecture about cyclic subgroups of G.

In this form a method of approach suggests itself, which, if it works for cyclic subgroups, will probably work for any subgroup. Thus let us try to prove that if H is any subgroup of G, then the order of H divides the order of G. The method then goes as follows. We have to partition G into (disjoint) subsets, each of which contains the same number of elements as does H.

The problem is to obtain some subsets T_i of G, such that

(1) $|T_i| = |H|$,

(2) these subsets are disjoint,

and (3) the union of these subsets is G.

To start us off, we certainly know one subset which has $|H|$ elements in it, namely H itself. With the information at our disposal, the obvious way of obtaining some more subsets T_i with $|T_i| = |H|$ is to multiply H by elements of G.

Thus with a fixed g from G, we consider sets like

$$Hg = \{hg \mid h \in H\} .$$

Since $\quad h_1 g = h_2 g \to h_1 g g^{-1} = h_2 g g^{-1} \to h_1 = h_2; \quad h_1, h_2 \in H,$

the number of distinct elements in Hg is $|H|$.

Since $g = eg \in Hg$, it follows that if we consider all sets of the form Hg, $g \in G$, the union of these sets is G.

However, the sets Hg, $g \in G$, are certainly not disjoint. Fortunately all is not lost.

Suppose we have 2 sets Hg, Hg' of the form being considered with $g, g' \in G$. Suppose $Hg \cap Hg' \neq \emptyset$. Let $x \in Hg \cap Hg'$. Then $x = h_1 g = h_2 g'$ for some $h_1, h_2 \in H$. But then $g = h_1^{-1} h_2 g' \in Hg'$. If $y = hg$ is <u>any</u> element of Hg, then

$$y = h(h_1^{-1} h_2) g' \in Hg' .$$

Thus $\quad Hg \subset Hg'$.

A similar argument with h_1, h_2 and g, g' interchanged shows that $Hg' \subset Hg$.

Hence altogether $Hg' = Hg$.

We conclude that any 2 sets of the form Hg, $g \in G$, are either disjoint or identical.

We now run through all sets of the form Hg, $g \in G$, and discard all repetitions. Let the sets retained be denoted by $Hg_1, Hg_2, \ldots, Hg_r$; where g_1 may be taken to be e, so that $Hg_1 = He = H$. Then we obtain the partition of G that we were looking for.

$$G = \bigcup_{i=1}^{r} Hg_i \quad ,$$

where $Hg_i \cap Hg_j = \emptyset$ when $i \neq j$.

Since we have partitioned G into r subsets each containing $|H|$ elements of G, we have the equation

$$|G| = r|H| \ .$$

We have proved:

THEOREM 3.1.1. *Let H be a subgroup of a finite group G, then $|H|$ divides $|G|$.*

In particular, if $g \in G$, then $|\langle g \rangle|$ divides $|G|$. Now $|\langle g \rangle| = O(g)$, hence $O(g) \mid |G|$.

We state this result as:

THEOREM 3.1.2. *Let g be an element of a finite group G then $O(g)$ divides $|G|$.*

In the literature one or other of these theorems is referred to as <u>**Lagrange's theorem**</u>. It varies from author to author.

As shown at the beginning of this section, we can assert now:

THEOREM 3.1.3. *Let G be a finite group of order a prime p. Then G is cyclic.*

As stated in section 2.7, we can now deduce the possible groups of orders 1, 2, 3 and 4, more easily than before. By Theorem 3.1.3 the groups of orders 1, 2, 3 are cyclic groups C_1, C_2, C_3, unique upto isomorphism. For a group G of order 4, either G has an element of order 4 and is isomorphic to the cyclic group C_4, or, by Theorem 3.1.2, all elements other than the identity e have order 2. Thus

$$G = \{e, a, b, c\}$$

with $a^2 = b^2 = c^2 = e$.

Hence $c = ab$, and $(ab)^2 = e$. Thus

$$ab = (ab)^{-1} = b^{-1}a^{-1} = ba.$$

Therefore

$$G = \langle a, b \mid a^2 = b^2 = e, ab = ba \rangle = K_4 \quad .$$

DEFINITION 3.1.1. *The sets Hg are usually called the* ***right cosets*** *of H in G. We could equally well have considered the* ***left cosets*** *of H in G, which are subsets of G of the form gH, $g \in G$. We have $|gH| = |H| = |Hg|$.*

Note. Some authors call the sets Hg, $g \in G$, left cosets and the sets gH, $g \in G$, right cosets.

3.2 Some Results on Subgroups

The idea of a subgroup was first mentioned in section 2.7. However, so far, we have said very little about this concept in a formal way. In this section we remedy this by collecting together several elementary results about subgroups of a group. For convenience we call this collection

THEOREM 3.2.1. *Let G be a group, possibly infinite.*

(1) A subset S of a group G is a subgroup (that is a group under the same product as in G) if and only if

(a) $a, b \in S \rightarrow ab \in S$.

(b) $e \in S$, where e is the identity of G

(c) $a \in S \rightarrow a^{-1} \in S$.

(2) More briefly, a non-empty subset S of a group G is a subgroup of G if and only if

$$a,\ b \in S \rightarrow ab^{-1} \in S\ .$$

(3) Even more simply, for a ***finite*** *group G, a non-empty subset S is a subgroup if and only if*

$$a,\ b \in S \rightarrow ab \in S\ .$$

(This means that for a ***finite*** *group it is enough that the product in G is a product for S.)*

(4) If $\{H_i\}$ is any set of subgroups of G, where $i \in \Delta$, some index set, then $\bigcap_{i \in \Delta} H_i$ is a subgroup of G.

PROOF

(1) If S is a subset for which the conditions hold, then the group axioms hold in S, associativity being inherited from G.

Conversely, if S is a subgroup of G, then (a) is immediate. If e' is the identity of S, then in S and in G we have: $e'e' = e'$. Hence $e' = e$, the identity of G. Thus $e \in S$. If s' is the inverse of s in S, then $e = e' = ss'$. Thus

$$s' = s^{-1}\ .$$

Hence

$$s^{-1} \in S\ .$$

This completes the proof of (1).

(2) If S is a subgroup of G, then by (1)

$$a,\ b \in S \rightarrow a,\ b^{-1} \in S \rightarrow ab^{-1} \in S\ .$$

Conversely, suppose $a,\ b \in S \rightarrow ab^{-1} \in S$. Then $a,\ a \in S \rightarrow aa^{-1} \in S \rightarrow e \in S$. Thus, since $S \neq \emptyset$, $e \in S$. If $a \in S$, we have e and a in S. Hence $ea^{-1} \in S$. Thus

$$a^{-1} \in S.$$

Finally, if $a,\ b \in S$, then by the above $b^{-1} \in S$. Hence

$$a,\ b^{-1} \in S.$$

Thus, by the condition of part (2),

$$a(b^{-1})^{-1} \in S\ .$$

Hence $ab \in S$. By part (1), S is a subgroup of G.

(3) If S is a subgroup, then S is closed under the product in G. Hence

$$a,\ b \in S \rightarrow ab \in S\ .$$

Conversely, if $a,\ b \in S \rightarrow ab \in S$, then (a) of part (1) is satisfied. If $a \in S$, then a^2, a^3, a^4, ... are all in S, by the given condition. Now G is finite, hence the order of a must be finite. Thus for some positive integer n, $a^n = e$. Thus

$$e = a^n \in S\ .$$

This is condition (b) of part (1).

Finally, if $a \in S$, then

$$a^{-1} = a^{n-1} \in S$$

for some n. This is condition (c) of part (1). Thus S is a subgroup of G.

(4) Let $a,\ b \in \bigcap_{i \in \Delta} H_i$. Then $a,\ b \in H_i$ for all $i \in \Delta$. Hence $ab^{-1} \in H_i$ for all $i \in \Delta$ because H_i is a subgroup. Thus,

$$ab^{-1} \in \bigcap_{i \in \Delta} H_i\ .$$

We have shown that

$$a,\ b \in \bigcap_{i \in \Delta} H_i \rightarrow ab^{-1} \in \bigcap_{i \in \Delta} H_i\ .$$

Thus, by part (2) above, $\bigcap_{i \in \Delta} H_i$ is a subgroup of G.

3.3 Generators

These were mentioned in section 2.6 in connection with the generators and relations description of a group. Here we give another definition similar to the idea in section 2.7 where the Klein 4-group was introduced as $\langle (12),\ (34) \rangle$, the subgroup of S_4 generated by (12), and (34). We make use of theorem 3.2.1 part (4).

Let S be a non-empty set of elements of G and let $\{H_i \mid i \in \Delta\}$ be the set of all subgroups H_i of G which contain S: $H_i \supset S$.

Then $\bigcap_{i \in \Delta} H_i$ is a subgroup of G by theorem 3.2.1 part (4).

DEFINITION 3.3.1. *The subgroup* $\bigcap_{i \in \Delta} H_i$ *is defined to be the subgroup of* G ***generated*** *by* S. S *is called a* ***generating set*** *of* $\bigcap_{i \in \Delta} H_i$. *The members of* S *are called* *generators* *of* $\bigcap_{i \in \Delta} H_i$. *If* S *is a finite set, then* $\bigcap_{i \in \Delta} H_i$ *is said to be finitely generated. We usually denote* $\bigcap_{i \in \Delta} H_i$ *by* $\langle S \rangle$.

As we know, if S has one member, $\langle S \rangle$ is a cyclic group.

The following theorem gives the explicit form of the members of $\langle S \rangle$.

THEOREM 3.3.1. *The members of* $\langle S \rangle$ *are* *precisely* *those elements of the form* $a_1 a_2 a_3 \ldots a_n$, *where either* $a_i \in S$ *or* $a_i^{-1} \in S$, *for* $i = 1, 2, 3, \ldots, n$.

PROOF Let $H = \{a_1 a_2 \ldots a_n \mid a_i \in S \text{ or } a_i^{-1} \in S\}$. Let $x, y \in H$.

Then
$$x = a_1 a_2 a_3 \ldots a_n,$$
$$y = b_1 b_2 \ldots b_m,$$
$$xy^{-1} = a_1 a_2 \ldots a_n b_m^{-1} \ldots b_2^{-1} b_1^{-1}.$$

Hence $xy^{-1} \in H$. Thus H is a subgroup, by theorem 3.2.1(2). Moreover $H \supset S$. Thus

$$H \supset \langle S \rangle ,$$

since $\langle S \rangle$ is the intersection of all subgroups like H which contain S. Now $\langle S \rangle$ is a subgroup containing S. Hence $\langle S \rangle$ contains every element of the form $a_1 a_2 a_3 \ldots a_n$, where $a_i \in S$ or $a_i^{-1} \in S$. Thus

$$\langle S \rangle \supset H .$$

From $H \supset \langle S \rangle$ and $\langle S \rangle \supset H$, we deduce

$$H = \langle S \rangle .$$

3.4 Products of Subsets of Groups

Cosets are products of elements of a group G and a given subgroup H. They proved useful in proving Lagrange's theorem. In a similar way more general products turn out to be important. Before looking more closely at the structure of cyclic groups, we dispose of a few fairly technical results about general products.

We consider $AB = \{ab \mid a \in A,\ b \in B\}$, where A and B are subsets of a group G. The product AB reduces to the left coset gH when $A = \{g\}$ and $B = H$. It gives the right coset Hg when $A = H$ and $B = \{g\}$.

An often used result is

THEOREM 3.4.1. *If S is any non-empty subset of the group G, then $SG = G = GS$.*

PROOF $SG = \{sg \mid s \in S,\ g \in G\}$. Now $s \in G$, thus $sg \in S$. Hence $SG \subset G$. Let $g \in G$ and $s \in S$, then $g = s(s^{-1}g) \in SG$. Thus $G \subset SG$.

The 2 inclusions imply that

$$G = SG.$$

Similarly $GS = G$.

Given the product AB, it is natural to ask if AB is a subgroup when both A and B are subgroups. The following theorem shows that AB is a subgroup only when certain conditions hold. Formally we have

THEOREM 3.4.2.

(1) If A and B are subgroups of a group G, then AB is itself a subgroup if and only if

$$AB = BA.$$

(2) When this is so, the group $\langle A, B\rangle$, generated by A and B, is equal to AB.

(3) $|AB| = |A||B|/|A \cap B|$ for A and B finite subgroups, where AB need not be a subgroup.

PROOF

(1) Suppose AB is a subgroup of G. Let $x \in BA$, then

$$x = ba = (b^{-1})^{-1}(a^{-1})^{-1} = (a^{-1}b^{-1})^{-1} .$$

Now $a^{-1}b^{-1} \in AB$, and, since AB is a subgroup, $(a^{-1}b^{-1})^{-1} \in AB$. Thus $x \in AB$. Hence $AB \supset BA$. Now let $x \in AB$. Then $x = y^{-1}$ for some $y \in AB$ because AB is a group. In fact we may take $y = x^{-1}$. Then

$$x = y^{-1} = (ab)^{-1} = b^{-1}a^{-1} \in BA,$$

where $y = ab$ because $y \in AB$. Hence $AB \subset BA$. Thus

$$AB = BA .$$

Conversely, suppose $AB = BA$, with A and B subgroups. Then noting that the associative law holds for products of subsets

$$(AB)(AB) = A(BA)B = A(AB)B = (AA)(BB) = AB,$$

where we have used $AA = A$ and $BB = B$ by theorem 3.4.1. Thus

$$(AB)(AB) = AB .$$

Let $x, y \in AB$. Now $y \in AB$, so $y = ab$ for some a and b. Hence

$$y^{-1} = (ab)^{-1} = b^{-1}a^{-1} \in BA .$$

However, $BA = AB$ by hypothesis. Hence $y^{-1} \in AB$. Thus

$$xy^{-1} \in (AB)(AB) = AB.$$

We have shown that

$$x, y \in AB \rightarrow xy^{-1} \in AB .$$

It follows by theorem 3.2.1 (2) that AB is a subgroup.

(2) $\langle A, B\rangle \supset AB$ because $\langle A, B\rangle$ contains all products like ab; $a \in A$, $b \in B$. Now $AB \supset A$ and $AB \supset B$. Thus $AB \supset \langle A, B\rangle$, when AB is a subgroup. Hence, when AB is a subgroup, we have

$$AB = \langle A, B\rangle .$$

(3) Put $D = A \cap D$. Now D is a subgroup of B so we can decompose B into right cosets relative to D as follows:

$$B = Db_1 \cup Db_2 \cup \ldots \cup Db_r ,$$

where the cosets are disjoint and $b_i \in B$; $i = 1, 2, \ldots, r$.

Then $\quad AB = ADb_1 \cup ADb_2 \cup \ldots \cup ADb_r$.

Now D is a subgroup of A. Hence

$$AD = A.$$

Thus $AB = Ab_1 \cup Ab_2 \cup \ldots \cup Ab_r$, which may not be disjoint. Suppose $x \in Ab_i \cap Ab_j$, $i \neq j$. Then

$$x = a_1 b_i = a_2 b_j$$

for some a_1, $a_2 \in A$. But then $b_i = a_1^{-1} a_2 b_j$. Hence

$$b_i b_j^{-1} = a_1^{-1} a_2.$$

Thus $\quad b_i b_j^{-1} \in A$.

Now $b_i b_j^{-1} \in B$. Hence

$$b_i b_j^{-1} \in D.$$

Thus $\quad Db_j = Db_i$, $\quad i \neq j$.

Now this contradicts the fact that the cosets Db_i; $i = 1, 2, 3, \ldots, r$; are disjoint. Hence

$$AB = Ab_1 \cup Ab_2 \cup \ldots \cup Ab_r ,$$

where the sets Ab_i are disjoint. Now $|Ab_i| = |A|$. Hence

$$|AB| = |A| + |A| + \ldots + |A| = r|A| .$$

Since $r = |B|/|D|$, we have:

$$|AB| = |B||A|/|D| .$$

Finally

$$|AB| = |A||B|/|A \cap B| .$$

Note. Part (3) of theorem 3.4.2 may also be proved quite neatly as follows.

If $x \in A \cap B$, then $ab = ax^{-1}xb = a'b'$, where $a' = ax^{-1} \in A$ and $b' = xb \in B$.

Thus each ab is repeated in the product AB at least $|A \cap B|$ times as a runs through the elements of A and b runs through the elements of B.

On the other hand, if $ab = a'b'$, then

$$b'b^{-1} = (a')^{-1}a \in A \cap B.$$

Put $b'b^{-1} = (a')^{-1}a = x \in A \cap B$. Then

$$b' = xb$$

and
$$a' = ax^{-1} .$$

Thus **all** repetitions are of the form given above. Hence each product ab is repeated **precisely** $|A \cap B|$ times. Thus $|AB| = |A||B|/|A \cap B|$.

3.5 Cyclic Groups

For cyclic groups the problem of classification mentioned in section 1.5 is solved by theorem 2.7.2, namely any infinite cyclic group is isomorphic to $(\mathbf{Z}, +)$ and any cyclic group of order n is isomorphic to $(\mathbf{Z}_n, +)$. However, there is another problem to be solved for cyclic groups, namely the structure problem. In particular we are interested in the subgroup structure; that is the nature of the subgroups of a cyclic group.

In general for an arbitrary group this is an unsolved problem. For cyclic groups there is a very satisfying solution, which we give as

THEOREM 3.5.1. *Every subgroup of a cyclic group* $\langle g \rangle$ *is cyclic. Moreover, if* $\langle g \rangle$ *is finite of order* n, *then there is just one cyclic subgroup of order* m *for each divisor* m *of* n. *This subgroup has the form* $\langle g^{n/m} \rangle$. *These are all the subgroups of* $G = \langle g \rangle$ *in the finite case.*

PROOF Let H be a subgroup of $\langle g \rangle$. If $H = \{e\}$, where e is the identity, then H is trivially cyclic. Let $h \neq e$ be any member of H. Then

$$h = g^s$$

for some s, and

$$h^{-1} = g^{-s} \in H .$$

Thus there are positive integers t such that $g^t \in H$.

Take the least such positive integer. Call it ℓ. Let h be any element of H. Then $h = g^s$ for some s. Write $s = q\ell + r$; $0 \leqq r < \ell$. Then $h = (g^\ell)^q . g^r$. Thus

$$g^r = (g^\ell)^{-q} h \in H.$$

If $r \neq 0$, then $r < \ell$, which contradicts the choice of ℓ. Thus $r = 0$. Then $h = (g^\ell)^q$. Thus

$$H \subset \langle g^\ell \rangle .$$

Now $g^\ell \in H$, hence

$$\langle g^\ell \rangle \subset H .$$

Thus $$H = \langle g^\ell \rangle .$$

Moreover if $\langle g \rangle$ is finite order n, then $g^n = e \in H$. Thus, taking $s = n$ in the above, $n = m\ell$ for some m. Hence

$$\ell = n/m$$

and the order of g^ℓ is m. Thus H is order m, where $H = \langle g^{n/m} \rangle$.

Now suppose we are given $m \mid n$. Then

$$\{g^{n/m}, g^{2n/m}, \ldots, g^{mn/m} = e\}$$

is a cyclic subgroup of order m. Hence to each divisor m of n, there corresponds just one subgroup $H = \langle g^{n/m} \rangle$ of order m. The above argument shows that these are all the subgroups of $G = \langle g \rangle$.

EXAMPLE. Take $G = \langle g \rangle$, cyclic order 24. Divisors of 24 are 1, 2, 3, 4, 6, 8, 12, 24. As usual let C_n denote a cyclic group of order n. To each divisor there is a corresponding cyclic subgroup. Hence the subgroups of G are:

$$C_1 = \{e\},\ C_2,\ C_3,\ C_4,\ C_6,\ C_8,\ C_{12},\ C_{24} = G.$$

These are often exhibited as points of a diagram called a lattice; in this case the **lattice of subgroups**. $C_n \searrow C_m$ means that C_m is a subgroup of C_n, and there is no subgroup of C_n containing C_m .

In our case we have the following lattice of subgroups.

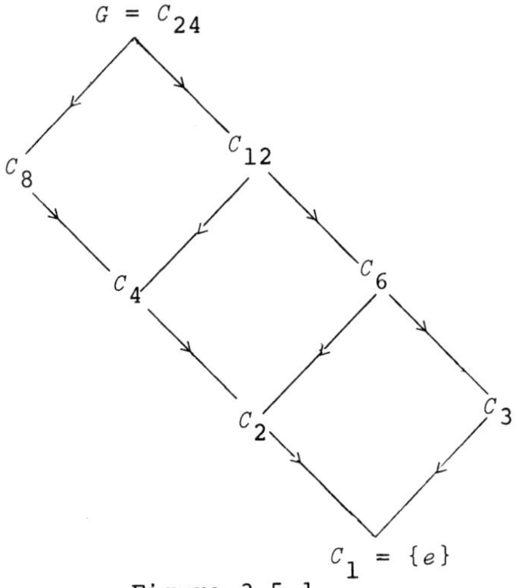

Figure 3.5.1

3.6 Subgroups of S_3

As a further example of the use of Lagrange's theorem in determining the structure of a group, let us find all the subgroups of S_3.

$$S_3 = \{(1), (12), (13), (23), (123), (132)\} .$$

$|S_3| = 6 = 2 \times 3$. By theorem 3.1.1 (Lagrange's theorem) the subgroups must have orders 1, 2, 3, 6.

The subgroup of order 1 must be $\{e\}$. The subgroup of order 6 must be S_3. By theorem 3.1.3, all subgroups of order 2 or 3 must be cyclic, generated by elements of order 2 or 3. This is sufficient to determine the other subgroups as:

$$\langle(12)\rangle, \langle(13)\rangle, \langle(23)\rangle, \langle(123)\rangle = \langle(132)\rangle .$$

Note. (123) and (132) generate the same subgroup, since

$$\langle(123)\rangle = \{(1), (123), (132)\} = \langle(132)\rangle .$$

The reader should try to find the subgroups of S_3 without the help of theorems 3.1.1 and 3.1.3 whose value will then be appreciated.

EXERCISES 3

1. Find an example of a group G that contains a subset S which is a group under a different multiplication from that in G, but is not a group under the multiplication in G.

2. Prove that every subgroup $\neq \{e\}$ of an infinite cyclic group G is isomorphic to G.

3. Let C_n denote a cyclic group of order n. Draw lattice diagrams of the subgroups of:

$$C_5, \; C_8, \; C_{12}, \; C_{18}, \; C_{25}, \; C_{36} \, .$$

Use a separate diagram for each group.

4. List all subgroups of $(\underline{\mathbf{Z}}_p, +)$ where p is prime.

5. Write out the multiplication table for the set G of integers 1, 2, 3, 4, 5, 6, with multiplication taken modulo 7. Hence enumerate the orders of all the elements. Show that this set G with this multiplication is a cyclic group of order 6.

6. For each subgroup H of S_3 find the right and left cosets of H in S_3.

7. Let

$$f = \begin{pmatrix} a_1 & a_2 & \dots & a_n \\ b_1 & b_2 & \dots & b_n \end{pmatrix} \in S_n \, ,$$

where the a_i and b_j are the numbers 1, 2, 3, 4, ..., n, in some order. Define a function $t: S_n \to \underline{\mathbf{Q}}$ (the field of rational numbers) by

$$t(f) = \Pi \left(\frac{a_i - a_j}{b_i - b_j} \right) \, ,$$

where the product Π is taken over all subsets $\{i, j\}$, $i \neq j$, of the set $\{1, 2, 3, 4, 5, \ldots, n\}$.

Calculate $t(f)$, when

$$f = \begin{pmatrix} 2 & 3 & 1 & 4 \\ 1 & 2 & 4 & 3 \end{pmatrix} .$$

8. Let the notation be as in question 6 above. Let

$$f = \begin{pmatrix} a_1 & a_2 & \cdots & a_n \\ b_1 & b_2 & \cdots & b_n \end{pmatrix}$$

and

$$g = \begin{pmatrix} b_1 & b_2 & \cdots & b_n \\ c_1 & c_2 & \cdots & c_n \end{pmatrix}$$

be elements of S_n. Prove that

$$t(gf) = t(g)t(f) .$$

9. Let $f \in S_n$ be a transposition (see exercise 2.9). Prove that $t(f) = -1$.

10. Let f be any permutation in S_n. Prove that $t(f) = +1$ if f is a product of an **even** number of transpositions and -1 if f is a product of an **odd** number of transpositions.

(**Note** that since $t(f)$ is determined by f this means that f **cannot** be expressed **both** as a product of an even number of transpositions and also as a product of an odd number of transpositions.)

11. If $t(f) = +1$ we say that f is an **even** permutation. If $t(f) = -1$ we say that f is an **odd** permutation. Let G be any subgroup of S_n. Let S be the set of **even** permutations in G. Prove that S is a subgroup of G.

(**Note** that if $G = S_n$, then S is called the **alternating group** on n symbols and is denoted by A_n. Also see exercise 4.15.)

12. Let $(a_1\ a_2\ a_3\ \ldots\ a_r)$ be a cycle of length r. Prove that

$$t((a_1\ a_2\ \ldots\ a_r)) = (-1)^{r-1} .$$

Hence find A_3.

CHAPTER 4.
FACTOR GROUPS, PERMUTATION REPRESENTATIONS, FINITE POINT GROUPS

4.1 Normal Subgroups

In order to prove the conjecture that 'the order of the subgroup divides the order of the group' in section 3.1, we introduced certain subsets of a group called right cosets. As pointed out in definition 3.1.1 we could equally well have considered left cosets. This leads at once to certain questions.

To pose these questions let us consider a group G and a subgroup H. Relative to these groups we have a collection of right cosets $\{Hg \mid g \in G\}$ and a collection of left cosets $\{g'H \mid g' \in G\}$. Each collection partitions the set G. As noted in definition 3.1.1, the number of elements in each coset left or right is $|H|$ the order of H.

Two questions suggest themselves.

(1) Are there the same number of left cosets as there are right cosets?

(2) Do the left cosets give the same partition of G as the right cosets? If not, when do they give the same partition?

The first question can be answered at once in the finite case. The number of right cosets $= |G|/|H|$. This only depends on the orders of G and H. Hence the number of left cosets must be $|G|/|H|$. This can be checked by going through the calculations for the left cosets in detail if desired. We conclude that there are the same number of left cosets as right cosets.

However, this method cannot be used in the case of an infinite group G. Let us state and prove the result in a form suitable for <u>any</u> group. This is

THEOREM 4.1.1. *There is a one-one correspondence between the set of all right cosets of H in G and the set of all left cosets of H in G. The common cardinality of the two sets is called the index of H in G and is usually denoted by $[G : H]$. In the finite case $[G : H] = |G|/|H|$. The number of elements in each coset is $|H|$.*

PROOF Let L be the set of left cosets and R be the set of right cosets. Define a function f: $L \to R$ by $f(gH) = Hg^{-1}$. This definition apparently depends on g. However, f is well-defined since if $g_1H = g_2H$, then $g_2^{-1}g_1 \in H$. Hence

$$(g_2^{-1})(g_1^{-1})^{-1} \in H.$$

Thus $\quad g_2^{-1} \in Hg_1^{-1}$.

Thus $\quad g_2^{-1} \in Hg_2^{-1} \cap Hg_1^{-1}$.

Hence $\quad Hg_2^{-1} = Hg_1^{-1}$,

because cosets are either identical or disjoint.

Therefore if $g_1H = g_2H$, then

$$f(g_1H) = f(g_2H) .$$

f is onto, since if Hg is any right coset we have

$$f(g^{-1}H) = H(g^{-1})^{-1} = Hg .$$

Finally f is one-one. To see this suppose that $f(g_1H) = f(g_2H)$. Then $Hg_1^{-1} = Hg_2^{-1}$. Thus

$$g_1^{-1} = hg_2^{-1}$$

for some $h \in H$. Hence

$$g_2 = g_1h .$$

Thus $\quad g_2 \in g_2H \cap g_1H$.

Now this means that $g_2H = g_1H$. It follows that f is one-one. The theorem is proved.

To answer question (2) let us consider a specific group, say S_3. We have already found the subgroups of S_3 in section 3.6. We take

$$S_3 = \{(1),\ (12),\ (13),\ (23),\ (123),\ (132)\}$$

and the subgroup $H = \{(1),\ (12)\}$.

The left cosets of H in S_3 are:

$$(1)\ \{(1),\ (12)\} = \{(1),\ (12)\}$$

$$(12)\ \{(1),\ (12)\} = \{(12),\ (1)\}$$

$$(13)\ \{(1),\ (12)\} = \{(13),\ (123)\}$$

$$(23)\ \{(1),\ (12)\} = \{(23),\ (132)\}$$

$$(123)\ \{(1),\ (12)\} = \{(123),\ (13)\}$$

$$(132)\ \{(1),\ (12)\} = \{(132),\ (23)\}$$

These give 3 distinct left cosets, namely

$$H = \{(1),\ (12)\},\ \{(13),\ (123)\},\ \{(23),\ (132)\} \quad .$$

The right cosets of H in S_3 are:

$$\{(1),\ (12)\}(1) \quad = \{(1),\ (12)\}$$

$$\{(1),\ (12)\}(12) \quad = \{(12),\ (1)\}$$

$$\{(1),\ (12)\}(13) \quad = \{(13),\ (132)\}$$

$$\{(1),\ (12)\}(23) \quad = \{(23),\ (123)\}$$

$$\{(1),\ (12)\}(123) = \{(123),\ (23)\}$$

$$\{(1),\ (12)\}(132) = \{(132),\ (13)\}$$

These give 3 distinct right cosets, namely

$$H = \{(1),\ (12)\},\ \{(13),\ (132)\},\ \{(23),\ (123)\} \ .$$

Hence the left and right cosets are <u>not</u> the same.

When do left and right cosets coincide? Suppose we have a group G and subgroup H. Since $g \in G$ can be written ge, where e is the identity of G, the left coset of H in G which contains the element g is gH. Similarly the right coset of H in G which contains g is Hg. It follows that the left and right cosets coincide if and only if

$$Hg = gH$$

for every $g \in G$.

A subgroup H having this property is called a **normal** subgroup of G although in many groups such subgroups are the exception rather than the rule!

We have the following:

THEOREM 4.1.2. *A subgroup H of a group G is* ***normal*** *in G if and only if any one of the following equivalent statements holds.*

(1) $gH = Hg$, *for all* $g \in G$,

(2) $g^{-1}Hg = H$, *for all* $g \in G$,

(3) $g^{-1}Hg \subset H$, *for all* $g \in G$,

(4) $g^{-1}hg \in H$, *for all* $g \in G$ *and all* $h \in H$.

PROOF In the following g and h are **arbitrary** elements, $h \in H$, $g \in G$.

(1) is the definition of a normal subgroup. We shall prove the chain of implication

$$(1) \to (2) \to (3) \to (4) \to (1)$$

from which the theorem follows.

(1) $\to$ (2): $gH = Hg \to g^{-1}gH = g^{-1}Hg \to H = g^{-1}Hg$

(2) $\to$ (3): $g^{-1}Hg = H \to g^{-1}Hg \subset H$

(3) $\to$ (4): $g^{-1}Hg \subset H \to g^{-1}hg \in H$; $h \in H$, $g \in G$

(4) $\to$ (1): $g^{-1}hg \in H \to g^{-1}Hg \subset H \to (g^{-1})^{-1}Hg^{-1} \subset H$ (taking g^{-1} for g) $\to gHg^{-1} \subset H \to g^{-1}gHg^{-1}g \subset g^{-1}Hg$.

Then $H \subset g^{-1}Hg$ and $g^{-1}Hg \subset H$ together give

$$g^{-1}Hg = H .$$

Then $$Hg = gH .$$

Thus (4) $\rightarrow$ (1). Altogether we have

$$(1) \rightarrow (2) \rightarrow (3) \rightarrow (4) \rightarrow (1) .$$

The theorem follows.

A useful result is the following:

THEOREM 4.1.3. *Let H be a subgroup of index 2 in the group G. Then H is normal in G. In particular A_n is normal in S_n.* (See exercise 3.11.)

PROOF Exercise 4.2 and exercise 4.15.

Note that any subgroup of an abelian group is normal.

4.2 Simplicity

DEFINITION 4.2.1. *A group G without normal subgroups other than $\{e\}$ and G, which are always normal, is called a* ***simple*** *group.*

The following theorems provide an infinite number of examples of simple groups.

THEOREM 4.2.1. *The abelian simple groups are precisely the cyclic groups of prime order.*

PROOF An abelian group G is simple if and only if its only subgroups are $\{e\}$ and G. Since any element of G generates a cyclic subgroup, it follows that G must be cyclic. From theorem 3.5.1 G must be of prime order.

THEOREM 4.2.2. *The alternating group, A_n, $n \geqq 5$, is a simple group.* (See exercise 3.11 for the definition of the alternating group.)

PROOF The proof breaks down into 3 distinct steps.

Step (1). We show that the cycles of length 3 generate A_n, the subgroup of S_n consisting of all the even permutations in S_n.

Every even permutation can be written as a product of an even number of transpositions. (See exercises 3.10 and 3.11.) Pairs of transpositions can be replaced by 3-cycles as follows:

$$(ik)(ij) = (ijk) ,$$

$$(k\ell)(ij) = (i\ell k)(ijk) ,$$

which covers the two possibilities. Hence every even permutation is a product of 3-cycles. Moreover every 3-cycle, being even, is in A_n. The required result follows.

Step (2). If H is a normal subgroup of A_n and the 3-cycle $(ijk) \in H$, then $H = A_n$.

To prove this statement, let $(i'j'k')$ be any other 3-cycle. Put

$$f = \begin{pmatrix} i & j & k & \ell & m & \dots \\ i' & j' & k' & \ell' & m' & \dots \end{pmatrix}$$

Since $n \geqq 5$, we have at least 5 symbols available. If f as written is odd, it can be made even by interchanging ℓ' and m' in the bottom row. Then

$$f(ijk)f^{-1} = (i'j'k') \ .$$

This follows by direct calculation. Hence $(i'j'k') \in H$ since H is normal. Thus H contains **all** 3-cycles. By step (1) the 3-cycles generate A_n. Hence $H \supset A_n$. Of course $A_n \supset H$. Thus

$$H = A_n \ ,$$

as required.

Step 3. We assert that if H is **any** normal subgroup of A_n, $n \geqq 5$, other than the identity subgroup $\{(1)\}$, then H contains at least one 3-cycle.

To prove this statement choose in H a permutation $f \neq (1)$, the identity permutation, such that f leaves fixed the largest possible number of symbols for any permutation in H. Since we are dealing with **finite** sets of permutations this is possible.

If f is **not** a 3-cycle, then it is, when expressed as a product of disjoint cycles, **either** of the form:

(a) $f = (ij)(k\ell)(\ \ )(\ \ ) \dots$

all cycles of length 2 and at least 2 such cycles in order to give an even permutation, **or** of the form:

(b) $f = (ijk \dots)(\dots) \dots$

where one cycle at least is of length $\geqq 3$. If there is only one cycle of length 3, then there must be some more cycles, because we are supposing that f is not a 3-cycle. Also f cannot be a 4-cycle on its own because a 4-cycle is odd. Hence f must involve at least 2 symbols other than i, j and k. Call these symbols ℓ and m.

Note. In case (a) above, let m be any symbol other than i, j, k and ℓ. Let $g = (k\ell m)$. Then

$$gfg^{-1} \in H ,$$

since $g \in A_n$, $f \in H$ and H is normal in A_n. Hence

$$f^{-1}gfg^{-1} \in H .$$

Put $h = f^{-1}gfg^{-1}$.

First we observe that $h \neq (1)$, because in case (a) gfg^{-1} takes ℓ into m, while f takes ℓ into k. Thus

$$f \neq gfg^{-1} .$$

Hence $(1) \neq f^{-1}gfg^{-1} = h$.

In case (b) gfg^{-1} takes j into ℓ, while f takes j into k. Thus again

$$f \neq gfg^{-1} .$$

Hence $(1) \neq f^{-1}gfg^{-1} = h$.

These statements are obtained by direct calculation.

Now in case (a) h leaves fixed all the symbols left fixed by f except possibly m. On the other hand h leaves fixed i and j. Thus on balance h leaves fixed at least one more symbol than does f.

In case (b) h leaves fixed all the symbols left fixed by f because these are distinct from k, ℓ, m. However, h leaves i fixed whereas f does not. Hence h leaves fixed more symbols than does f.

Thus in either case we get a contradiction to the choice of f.

We conclude that f must be a 3-cycle on its own. Thus H contains at least one 3-cycle.

Steps (2) and (3) together imply that A_n is a simple group. This concludes the proof.

Note. We shall show in Chapter 8 that simple groups are the 'building blocks' from which all finite groups can be constructed.

4.3 Conjugacy

By theorem 4.1.2 (2), a subgroup H is normal in G if and only if $g^{-1}Hg = H$. This suggests looking more closely at subsets of G of the form

$$g^{-1}Mg = \{g^{-1}mg \mid m \in M\} ,$$

where M is a subset but not necessarily a subgroup of G.

First we note:

THEOREM 4.3.1. *If M is a subset of the group G, then the subset $g^{-1}Mg$ of G is a subgroup of G, if and only if M is a subgroup of G. When M is a subgroup, $g^{-1}Mg \cong M$. In any case $|g^{-1}Mg| = |M|$.*

PROOF Let M be a subgroup of G. Let $x = g^{-1}m_1g$ and $y = g^{-1}m_2g$ be any two elements of $g^{-1}Mg$. Then

$$xy^{-1} = (g^{-1}m_1g)(g^{-1}m_2g)^{-1} = g^{-1}m_1g.g^{-1}m_2^{-1}g$$

$$= g^{-1}m_1m_2^{-1}g \in g^{-1}Mg ,$$

because $m_1m_2^{-1} \in M$ as M is a subgroup. Thus

$$xy^{-1} \in g^{-1}Mg .$$

By theorem 3.2.1 (2), $g^{-1}Mg$ is a subgroup of G.

Conversely suppose $g^{-1}Mg$ is a subgroup of G. Let $N = g^{-1}Mg$. Then N is a subgroup of G. Hence, by the above,

$$M = gNg^{-1} = (g^{-1})^{-1}N(g^{-1}) ,$$

is also a subgroup of G.

Define the function $f\colon M \to g^{-1}Mg$ by $f(m) = g^{-1}mg$; $m \in M$. We leave the reader to check that f is an isomorphism of M with $g^{-1}Mg$.

Suppose $g^{-1}m_1g = g^{-1}m_2g$, then $g(g^{-1}m_1g)g^{-1} = g(g^{-1}m_2g)g^{-1}$. Hence

$$m_1 = m_2; \quad m_1, m_2 \in M .$$

Thus in any case $|g^{-1}Mg| = |M|$.

DEFINITION 4.3.1. *$g^{-1}Mg$ is called a* ***conjugate set*** *to M in G.*

As we let g run through G, we get a collection of sets conjugate to M.

By theorem 4.1.2 (2), H is a normal subgroup of G if and only if H is self-conjugate; that is H is the only set conjugate to H in G. Thus for a normal subgroup H there is only one set conjugate to H in G, namely itself.

In general, if M is **any** set in G, how many **distinct** sets are there conjugate to M in G ?

First we note that if $y^{-1}My = M$, then all elements of G of the form yx will give the same conjugate set, namely:

$$(yx)^{-1}M(yx) = x^{-1}y^{-1}Myx = x^{-1}Mx \quad .$$

Conversely if g and x lead to the same conjugate set, namely:

$$g^{-1}Mg = x^{-1}Mx ,$$

then

$$xg^{-1}Mgx^{-1} = M .$$

Thus if we put $y = gx^{-1}$, then

$$y^{-1}My = M$$

and

$$g = yx .$$

This means that if we put

$$N = \{y \in G \mid yM = My\} = \{y \in G \mid y^{-1}My = M\} ,$$

then the set of elements of the form Nx, $x \in G$, all lead to the same conjugate set $x^{-1}Mx$. Conversely if g gives the same conjugate as x, then $g \in Nx$.

Thus the number of distinct conjugate sets to M is equal to the number of distinct sets of the form Nx, as x runs through G. This number would be the index of N in G, by theorem 4.1.1, **if**

N were to be a **subgroup** of G. Let us, therefore, see if this is so.

Take any elements $y_1, y_2 \in N$. Then

$$y_2 M = M y_2 \rightarrow M = y_2^{-1} M y_2 \rightarrow M y_2^{-1} = y_2^{-1} M .$$

Thus $\quad y_1 y_2^{-1} M = y_1 M y_2^{-1} = M y_1 y_2^{-1}.$

Hence $y_1 y_2^{-1} \in N$. By theorem 3.2.1 (2) N is a subgroup of G. We call N the normaliser of M in G and usually write

$$N = N_G(M) .$$

Formally we have:

DEFINITION 4.3.2. *Let M be a subset of the group G. The set*

$$N_G(M) = \{y \in G \mid yM = My\} ,$$

is called the normaliser of M in G. If the set M consists of just one element m, then $N_G(m)$ is often called the centraliser of m in G, and is denoted by $C_G(m)$.

In the above argument we have proved:

THEOREM 4.3.2. *Let M be a subset of a group G. Then the number of distinct sets conjugate to the set M in G is $[G : N_G(M)]$.*

Easy but useful results are:

THEOREM 4.3.3. *Let G be a group.*

(1) Conjugate subgroups are isomorphic; in particular they have the same order.

(2) If H is a subgroup of G, then H is normal in $N_G(H)$.

(3) If N is <u>any</u> subgroup of G such that H is normal in N, then $H \subset N \subset N_G(H)$.

PROOF Exercise 4.3.

4.4 Conjugacy Classes

Of particular interest is the special case when M is a one element set. Let $M = \{m\}$, $m \in G$. It is usual to write

$$N_G(M) = N_G(\{m\}) ,$$

just as $N_G(m)$ or $C_G(m)$, and to talk about the normaliser or centraliser of m in G. According to theorem 4.3.2 the number of elements y of G which are conjugate to x in G is $[G : N_G(x)]$.

Let us define a relation R on G according to the following condition. Take $x, y \in G$, and define xRy if and only if $y = g^{-1}xg$, for some $g \in G$. A routine check shows that R is an equivalence relation (see [G]) on G. The equivalence classes are called **conjugacy classes** of G. The conjugacy class containing x consists precisely of all those elements y of G, which are conjugate to x; that is $y = g^{-1}xg$ for some $g \in G$. The number of elements of G in the conjugacy class containing x is $[G : N_G(x)]$ as remarked above.

THEOREM 4.4.1. *Let G be a group.*

(1) All elements in the conjugacy class containing x have the same order.

(2) H is normal in G if and only if H consists of the union of complete conjugacy classes.

PROOF Exercise 4.14.

When is the element $x \in G$ alone in its conjugacy class? It is an exceptional element. Hence the study of these elements is quite likely to lead to interesting results.

Now x is alone in its conjugacy class if and only if $x = g^{-1}xg$ for all $g \in G$. The collection of all elements with this property is called the **centre** of the group. This turns out to be one of the most important subgroups of a group. For future reference we give

DEFINITION 4.4.1. *The centre $Z(G)$ of a group G is defined by*

$$Z(G) = \{x \in G \mid xg = gx \quad \text{for all} \quad g \in G\} \quad .$$

We collect together several useful results about the centre of a group as

THEOREM 4.4.2. *Let G be a group. Let $Z(G)$ be the centre of G.*

(1) $Z(G)$ is a normal abelian subgroup of G.

(2) Let G be a ***finite*** *group. We have the* ***class equation***

$$|G| = |Z(G)| + h_1 + h_2 + \ldots + h_r \ ; \quad h_i > 1 \ ,$$

where h_i = *the number of elements in the* i^{th} *conjugacy class* = $[G : N_G(x_i)]$ *for any element* x_i *in the* i^{th} *conjugacy class.*

(3) If $|G| = p^r$ *for some prime* p *(G is often then called a* ***p-group*** *or a group of* ***prime power order*** *or a* ***prime power group****), then*

$$|Z(G)| = p^s > 1 .$$

In particular $Z(G) \neq \{e\}$ *for a p-group G.*

PROOF Let $Z(G)$ be the centre of the group G.

(1) Straightforward applications of theorems 3.2.1 (2) and 4.1.2 (4) suffice to show that $Z(G)$ is a normal subgroup of G. It is immediate that $Z(G)$ is abelian.

(2) Because the conjugacy classes form a partition of G, it follows that

$$|G| = \underbrace{1 + 1 + 1 + \ldots + 1}_{t \text{ terms}} + h_1 + h_2 + \ldots + h_r .$$

The t conjugacy classes each consisting of just one element of G provide us with the t 1's above. However, these are precisely the central elements. Thus

$$|Z(G)| = t .$$

The class equation follows.

(3) If $|G| = p^r$ for some prime p, then the class equation gives

$$p^r = |Z(G)| + h_1 + h_2 + \ldots + h_r ;$$

with $h_i > 1$. Now

$$h_i = [G : N_G(x_i)] = |G|/|N_G(x_i)| .$$

Thus h_i divides $|G|$; and $h_i > 1$. Thus $h_i = p^{\ell_i}$ for some $\ell_i \geqq 1$. Rewrite the class equation in the form

$$\begin{aligned} |Z(G)| &= p^r - h_1 - \ldots - h_r \\ &= p^r - p^{\ell_1} - p^{\ell_2} - \ldots - p^{\ell_r} ; \quad r \geqq 1,\ \ell_i \geqq 1. \end{aligned}$$

Then p divides $|Z(G)|$. Since $Z(G)$ is a subgroup of G, Lagrange's theorem 3.1.1 assures us that $|Z(G)|$ divides $|G|$. It follows from these two results that

$$|Z(G)| = p^s > 1 .$$

Thus a finite p-group has a non-trivial centre, i.e. $Z(G) \neq \{e\}$.

EXAMPLE 4.4.1. Let us pause to see how these ideas work out in the particular case of the group S_3 of order 6. Our work so far shows all groups of orders 1, 2, 3, 4, 5, to be abelian. Thus S_3 is the smallest non-abelian group. It is therefore the simplest group which gives non-trivial examples of most of the ideas that we have discussed or will discuss.

$$S_3 = \{(1), (12), (13), (23), (123), (132)\} .$$

(1) is in $Z(S_3)$ so is the only element in its conjugacy class.

$$(12)(13) = (132) , \qquad (13)(12) = (123)$$

$$(12)(23) = (123) , \qquad (23)(12) = (132) .$$

Thus not one of (12), (13), (23) lies in $Z(S_3)$. Hence they cannot be on their own in their respective conjugacy classes. By theorem 4.4.1 they, of order 2, cannot be conjugate to (123), or (132), because the latter have order 3. It follows that $\{(12), (13), (23)\}$ forms one conjugacy class.

$$(123)(12) = (13), \quad (12)(123) = (23)$$

hence $(123) \notin Z(S_3)$. Thus (123) is not on its own in its conjugacy class. Since (132) is the only other element of order 3 for it to be conjugate to, we must have the conjugacy class $\{(123), (132)\}$.

The conjugacy classes of S_3 are therefore:

$$\{(1)\}, \{(12), (13), (23)\}, \{(123), (132)\} .$$

Of course, we could obtain these by explicit methods such as:

$$(12)^{-1}(13)(12) = (12)(13)(12) = (23) .$$

Thus (23) is conjugate to (13); and so on.

The class equation is:

$$|S_3| = |Z(S_3)| + h_1 + h_2 ,$$

where h_1 is the number of elements in the conjugacy class $\{(12), (13), (23)\}$ and h_2 is the number of elements in the conjugacy class $\{(123), (132)\}$. We have

$$h_1 = [S_3 : N_{S_3}((12))] = |S_3|/|N_{S_3}((12))| .$$

Now $x \in N_{S_3}((12))$ if and only if $x(12) = (12)x$, if and only if $x = (1)$ or $x = (12)$. Thus

$$N_{S_3}((12)) = \{(1), (12)\} .$$

Hence $\quad h_1 = |S_3|/|N_{S_3}((12))| = 6/2 = 3 .$

Similarly $\quad h_2 = |S_3|/|N_{S_3}((123))| = |S_3|/|\{(1), (123), (132)\}|$

$$= \frac{6}{3}$$
$$= 2 .$$

The class equation is then

$$6 = 1 + 3 + 2.$$

Note that the centre of S_3 is trivial, i.e. $\{(1)\}$. In fact S_n has a trivial centre for all $n > 2$. We shall not prove that here.

EXAMPLE 4.4.2. Take as another example S_5 and consider $f = (12)(345)$. We calculate the number of permutations conjugate to f. Using the result in exercise 4.7, this number is 20. According to theorem 4.4.2 (2) this number is also equal to $[S_5 : N_{S_5}(f)]$. Thus the order of $N_{S_5}(f)$ is

$$\frac{5!}{20} = \frac{120}{20} = 6 .$$

Hence we have found a specific subgroup of order 6 of S_5. It consists of all permutations of S_5 which commute with f. In fact it is fairly easy to list the elements of $N_{S_5}(f)$ explicitly. $N_{S_5}(t)$ is bound to contain the cyclic subgroup generated by f. Now f is of order the L.C.M. of 2 and 3; that is 6. Thus $N_{S_5}(f)$, being of order 6, is the cyclic group generated by f. Working this out we get

$N_{S_5}(f) = \{(1), f, f^2 = (345)^2, f^3 = (12), f^4 = (345), f^5 = (12)(345)^2\}$.

4.5 Homomorphisms

In section 2.5 we discussed the idea of isomorphism and gave a definition in 2.5.1. If we can show that the group in which we are interested is isomorphic to a group whose structure is known, then our group can be identified with the known group. Hence its structure is known also. However, it is not always possible to obtain an isomorphism. Usually we have to be content with a function, f from our group into a known group, which is not one-one and onto but still preserves the structure in the sense that $f(g_1g_2) = f(g_1)f(g_2)$.

Formally we have:

DEFINITION 4.5.1. *Let G and H be groups. Let f be a function from G into H. If f satisfies the condition* $f(g_1g_2) = f(g_1)f(g_2)$ *for all* $g_1, g_2 \in G$, *then f is called a* ***homomorphism*** *from G into H. If f is* ***one-one*** *it is often called a* ***monomorphism***. *If f is* ***onto*** *it is often called an* ***epimorphism***. *An isomorphism is a monomorphism onto.*

Certain properties of homomorphisms follow at once. We collect these as:

THEOREM 4.5.1. *Let* e_G *be the identity of G. Let* e_H *be the identity of H. Let f be a homomorphism from G into H. Then*

(1) $f(e_G) = e_H$

(2) $f(g^{-1}) = [f(g)]^{-1}$, *for all* $g \in G$.

(3) Image of f = *Imf is a subgroup of H.*

PROOF

(1) Consider $f(g) = f(ge_G) = f(g)f(e_G)$.

Thus $[f(g)]^{-1}f(g) = [f(g)]^{-1}f(g)f(e_G)$.

Hence $e_H = f(e_G)$.

(2) $e_H = f(e_G) = f(gg^{-1}) = f(g)f(g^{-1})$.

Thus $[f(g)]^{-1}e_H = [f(g)]^{-1}f(g)f(g^{-1})$.

Hence $[f(g)]^{-1} = e_H f(g^{-1}) = f(g^{-1})$.

(3) Take $h_1, h_2 \in \operatorname{Im} f$, then $f(g_1) = h_1$, $f(g_2) = h_2$, for some $g_1, g_2 \in G$. Then $f(g_1 g_2^{-1}) = f(g_1)f(g_2^{-1}) = f(g_1)[f(g_2)]^{-1}$, by (2).

Thus $f(g_1 g_2^{-1}) = h_1 h_2^{-1}$.

Hence $h_1 h_2^{-1} \in \operatorname{Im} f$.

By theorem 3.2.1 (2), $\operatorname{Im} f$ is a subgroup of H. This concludes the proof.

Partition of G given by a homomorphism f.

Let f be a homomorphism from G into H. There is a natural partition of G produced by f, namely the subset $\bar{x}$ of G containing $x \in G$ consists of precisely those elements $g \in G$ which map onto the same element of H as does x. That is

$$g \in \bar{x} \leftrightarrow f(g) = f(x).$$

In particular $\bar{e}_G = \{x \in G \mid f(x) = e_H\}$, by theorem 4.5.1 (1). We usually drop the suffices G and H, writing e for e_G and e_H.

DEFINITION 4.5.2. $\bar{e}_G = \{x \in G \mid f(x) = e_H\}$ *is usually called the kernel of f, and abbreviated by* $\operatorname{Ker} f$.

We have

THEOREM 4.5.2. *The kernel of f, $\operatorname{Ker} f$, is a normal subgroup of G; where $f : G \to H$ is a homomorphism.*

PROOF Let $x, y \in \operatorname{Ker} f$. Then $f(x) = f(y) = e$. Then

$$f(x)[f(y)]^{-1} = e .$$

Thus $f(x)f(y^{-1}) = e$,

by theorem 4.5.1 (2). Hence

$$f(xy^{-1}) = e .$$

Thus $xy^{-1} \in \operatorname{Ker} f$. By theorem 3.2.1 (2), $\operatorname{Ker} f$ is a subgroup of G.

Let $x \in \operatorname{Ker} f$, $g \in G$. Then

$$f(g^{-1}xg) = f(g^{-1})f(x)f(g) = [f(g)]^{-1}f(x)f(g)$$

$$= f(g)^{-1}ef(g) = e .$$

Thus $g^{-1}xg \in \operatorname{Ker} f$. By theorem 4.1.2 (4), $\operatorname{Ker} f$ is normal in G. This concludes the proof.

Now each subset $\bar{x}$ of this partition is associated with just one element, namely $f(x)$ of $\operatorname{Im} f$. This suggests that from the homomorphism f we might be able to get an ismorphism $\bar{f}$, say, between $\operatorname{Im} f$ and some group obtained from G. In this way, although G itself is not isomorphic to $\operatorname{Im} f$, some group closely related to G might be isomorphic to $\operatorname{Im} f$. Hence, knowing the structure of $\operatorname{Im} f$ we would know also the structure of a group closely related to G, with the hope that this would lead to information about G itself.

Let us call the set $\bar{G}$, whose elements are the subsets $\bar{x}$, the quotient or factor set of G. The elements $\bar{x}$ of G are in one-one correspondence $\bar{f}$ with the elements of $\operatorname{Im} f$. We have $\bar{f}$ defined by:

$$\bar{f}(\bar{x}) = f(x) ,$$

where $\bar{x}$ is the subset of G containing x in the partition of G given by f.

For $\bar{G}$ to be a group with $\bar{f}$ an isomorphism of $\bar{G}$ with $\operatorname{Im} f$, the multiplication in $\bar{G}$ is determined as follows.

Go over to the elements $f(x)$ and $f(y)$ corresponding to $\bar{x}$ and $\bar{y}$ in $\operatorname{Im} f$. Multiply $f(x)$ and $f(y)$ in $\operatorname{Im} f$ to get

$$f(x).f(y) = f(xy) .$$

Then the product of $\bar{x}$ and $\bar{y}$ in $\bar{G}$ is the element in $\bar{G}$ corresponding to $f(xy)$. Now this is $\overline{(xy)}$. Hence the required product in $\bar{G}$ is

$$\bar{x}.\bar{y} = \overline{(xy)} .$$

It is instructive to describe the subsets $\bar{x}$ in another way. By theorem 4.5.2, $\operatorname{Ker} f = K$, say, is a normal subgroup of G. Let us consider the partition of G into cosets of K in G. Since K is normal it does not matter whether we consider left or right cosets. The coset containing x is xK. How is this set related to $\bar{x}$ considered above?

If $y \in xK$, then $y = xk$ for some $k \in \operatorname{Ker} f$. Thus

$$f(y) = f(xk) = f(x)f(k) = f(x)e_H ,$$

since $k \in \operatorname{Ker} f$. Hence

$$f(y) = f(x) .$$

This means that $y \in \bar{x}$. Thus $xK \subset \bar{x}$.

Conversely, if $y \in \bar{x}$, then $f(y) = f(x)$, from the definition of $\bar{x}$. Then

$$f(x^{-1}y) = f(x^{-1})f(y) = [f(x)]^{-1}f(y) = [f(x)]^{-1}f(x) = e_H .$$

Thus $x^{-1}y \in K$. Hence $y \in xK$.

Now y is any element of $\bar{x}$, therefore $\bar{x} \subset xK$. Together with the reverse inequality obtained earlier, we have

$$\bar{x} = xK .$$

The quotient or factor group $\bar{G} \cong \operatorname{Im} f$ is therefore just the group whose elements are the cosets xK of K in G with multiplication

$$xK.yK = xyK .$$

Put in this form we can see that given <u>**any**</u> normal subgroup N of the group G we might be able to construct a 'factor group' whose elements are the cosets xN of N in G with multiplication

(a) $$xN.yN = xyN .$$

First we need to check that the above equation (a) defines a valid product. As it stands the product could depend on the x and y chosen to specify the cosets xN and yN. However, if we consider xN and yN as subsets of G then the set

$$xNyN = \{xn_1yn_2 \mid n_1, n_2 \in N\}$$

of elements of G is also equal to

$$xyNN = xyN .$$

This is so because for any subgroup N, we have

$$NN = \{n_1 n_2 \mid n_1, n_2 \in N\}$$

is equal to N, and

$$yN = Ny$$

since N is normal. Thus

$$xNyN = xyN .$$

Hence any element from xN multiplied by any element from yN is an element of xyN. Thus if $xN = x'N$ and $yN = y'N$; that is if we choose x' and y' instead of x and y to represent the cosets xN and yN, then $x'N.y'N = x'y'N$ using (a). However, $x'y' \in xyN$. Thus

$$x'y'N = xyN .$$

Hence the definition of product given by (a) does not depend on the choice of representative elements in the cosets concerned.

It is a routine matter to check that the set of cosets with product defined by (a) is a group. Associativity follows from

$$xN.(yN.zN) = x(yz)N = (xy)zN = (xN.yN).zN .$$

The identity is $e_G N = N$, since

$$N.xN = xN = xN.N .$$

The inverse of xN is $x^{-1}N$, since

$$xN.x^{-1}N = xx^{-1}N = e_G N = N$$

and $$x^{-1}N.xN = x^{-1}xN = e_G N = N .$$

We call the set of cosets, with the product given by (a), the factor (or quotient) group of G relative to the normal subgroup N. We denote this group by $\frac{G}{N}$.

Formally we have:

DEFINITION 4.5.3. *Let G be a group. Let N be a* ***normal*** *subgroup of G. Let $\frac{G}{N}$ denote the set of cosets of N in G. Define a product on the set $\frac{G}{N}$ according to $xN.yN = xyN$. Relative to this product $\frac{G}{N}$ is a group called the* ***factor*** *or* ***quotient group*** *of G relative to the* ***normal*** *subgroup N.*

The factor group associated with the homomorphism f is $\frac{G}{K}$, where $K = \text{Ker}\, f$.

The results obtained earlier are summed up in the first part of the following theorem.

THEOREM 4.5.3.

(1) Let f be a homomorphism from the group G into the group H. Let $K = \text{Ker}\, f$. Let $\bar{x} = xK$, the coset containing x. Then $\bar{f}$ defined by $\bar{f}(\bar{x}) = f(x)$ is an isomorphism of $\bar{G} = \frac{G}{K}$ with $\text{Im}\, f$. $\frac{G}{K} \cong \text{Im}\, f$.

(2) Conversely if N is a normal subgroup of the group G then there is a natural epimorphism ℓ from G onto the factor group $\frac{G}{N}$.

ℓ is given by $\ell(g) = gN$; $g \in G$. The kernel of ℓ is N.

(3) Let f be a homomorphism from a group G into a group H. Let $K = \text{Ker}\, f$. Then we have the commutative diagram:

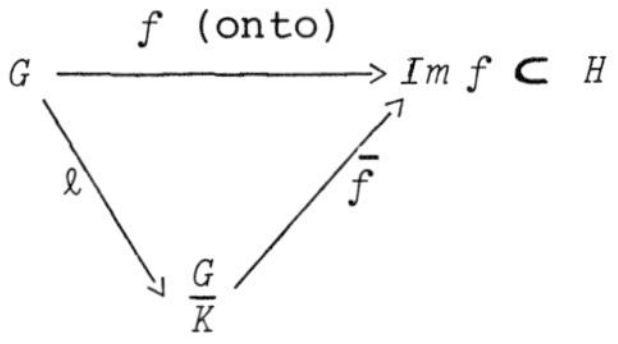

Figure 4.5.1

Note. *By commutative we mean that $f = \bar{f} \circ \ell$. Here $\bar{f}$ is the isomorphism obtained from f, and ℓ is the natural epimorphism of part (2), above.*

PROOF

(1) Already proved above.

(2) Let $\bar{g} = gN$. Given $\bar{g} \in \frac{G}{N}$, then $\ell(g) = \bar{g}$. Hence ℓ is onto.

Also $\qquad \ell(g_1 g_2) = \overline{(g_1 g_2)} = \bar{g}_1 \cdot \bar{g}_2 = \ell(g_1)\ell(g_2)$.

Thus ℓ is an epimorphism.

Note that the identity of $\frac{G}{N}$ is $\bar{e} = eN = N$. Let $g \in \text{Ker}\, \ell$. Then

$$\ell(g) = \bar{e} = N.$$

Now $\ell(g) = gN$. Thus

$$gN = N .$$

Hence $g \in N$. Thus Ker $\ell \subset N$.

Conversely, if $g \in N$, then $\ell(g) = gN = N = \bar{e}$. Thus $g \in$ Ker ℓ . Hence

$$N \subset \text{Ker } \ell .$$

Altogether we have Ker $\ell = N$.

(3) By part (1) we have $\bar{f} : \frac{G}{K} \cong \text{Im} f$. We have to check that $f = \bar{f} \circ \ell$. Let $g \in G$. Then

$$(\bar{f} \circ \ell)(g) = \bar{f}(\ell(g)) = \bar{f}(\bar{g}) = f(g) ,$$

by the definition of $\bar{f}$. Thus $\bar{f} \circ \ell = f$, as required.

DEFINITION 4.5.4. *Let G be a group. Let $a, b \in G$. The element $a^{-1}b^{-1}ab$ written $[a, b]$ is called the* ***commutator*** *of the elements a, b.*

In terms of commutators we can define a particularly important normal subgroup of G.

DEFINITION 4.5.5.

$$G' = \langle \{a^{-1}b^{-1}ab \mid a, b \in G\} \rangle$$

is called the ***commutator*** *or* ***derived subgroup*** *of G.*

Its importance comes from

THEOREM 4.5.4. *Let G be a group. Let G' be the commutator subgroup of G. Then*

(1) G' is normal in G.

(2) $\frac{G}{G'}$ is abelian.

(3) If $\frac{G}{N}$ is abelian, then $G' \subset N$.

PROOF Let G be a group and let G' be its derived subgroup.

(1) Let $x \in G'$, $g \in G$. Then

$$x = y_1 y_2 \dots y_r ,$$

where $y_i = [a_i, b_i]$; $a_i, b_i \in G$; $i = 1, 2, \dots, r$. Thus

$$g^{-1}xg = g^{-1}y_1 g \cdot g^{-1}y_2 g \cdot \dots \cdot g^{-1}y_r g .$$

Now $g^{-1}y_ig = g^{-1}[a_i, b_i] = g^{-1}a_i^{-1}b_i^{-1}a_ib_ig$

$$= g^{-1}a_i^{-1}g.g^{-1}b_ig.g^{-1}a_ig.g^{-1}b_ig$$

$$= (g^{-1}a_ig)^{-1}(g^{-1}b_ig)^{-1}(g^{-1}a_ig)(g^{-1}b_ig)$$

$$= [g^{-1}a_ig, g^{-1}b_ig] \in G' \qquad \text{for } i = 1, 2, \ldots,$$

Thus $g^{-1}xg \in G'$. By theorem 4.1.2 (4), G' is normal in G.

(2) Let $\bar{g}$ denote $gG' \in \frac{G}{G'}$. Let $\bar{g}_1, \bar{g}_2 \in \frac{G}{G'}$. Now

$$g_2g_1 = g_1g_2(g_2^{-1}g_1^{-1}g_2g_1) = g_1g_2[g_2, g_1].$$

Thus $g_2g_1 \in g_1g_2G'$. Hence

$$g_2g_1G' = g_1g_2G' .$$

That is $\overline{(g_2g_1)} = \overline{(g_1g_2)}$.

Thus $\bar{g}_1.\bar{g}_2 = \overline{(g_1g_2)} = \overline{(g_2g_1)} = \bar{g}_2.\bar{g}_1$.

Hence $\frac{G}{G'}$ is abelian.

(3) Let $\frac{G}{N}$ be abelian. Let $a, b \in G$. Let $\bar{a} = aN$, $\bar{b} = bN$.

Then $\bar{a}.\bar{b} = \bar{b}.\bar{a}$ because $\frac{G}{N}$ is abelian. Hence

$$\overline{(ab)} = \overline{(ba)} .$$

Thus $abN = baN$.

Hence $a^{-1}b^{-1}abN = N$.

Thus $[a, b] \in N$. This means that all the generators (see definition 3.3.1) of G' are elements of N. Hence $G' \subset N$. This concludes the proof of the theorem.

4.6 Permutation Representation of a Group

We digress from the build up of the ideas required for the pursuit of our main aim in order to illustrate some of the ideas discussed recently. We do this by introducing the concept of a permutation representation of a group. This is of interest in itself besides being of importance in many applications.

DEFINITION 4.6.1. *Any homomorphism of a group G into the symmetric group S_n is called a permutation representation of G.*

A particularly important permutation representation is obtained as follows.

Let H be an arbitrary subgroup of G. Let S_n be the symmetric group of all permutations on the n left cosets of H relative to G, where $[G : H] = n$. Define a function $f ; G \to S_n$ by $f(g) = P_g$ where P_g is the permutation on the n left cosets $x_1H, x_2H, \ldots, x_nH$; $x_i \in G$; given by

$$P_g = \begin{pmatrix} x_1H & x_2H & \ldots & x_nH \\ gx_1H & gx_2H & \ldots & gx_nH \end{pmatrix}$$

i.e. $P_g(x_iH) = gx_iH$. Now

$$f(g_1g_2) = P_{g_1g_2} = \begin{pmatrix} x_1H & \ldots & x_nH \\ g_1g_2x_1H & \ldots & g_1g_2x_nH \end{pmatrix}$$

$$= \begin{pmatrix} g_2x_1H & \ldots & g_2x_nH \\ g_1g_2x_1H & \ldots & g_1g_2x_nH \end{pmatrix} \begin{pmatrix} x_1H & \ldots & x_nH \\ g_2x_1H & \ldots & g_2x_nH \end{pmatrix}$$

$$= P_{g_1}P_{g_2} = f(g_1)f(g_2).$$

Hence f is a homomorphism. Thus we have obtained a permutation representation of G.

We have the following result concerning the kernel of f.

THEOREM 4.6.1. *Let N be any normal subgroup of G such that $N \subset H$. Then $N \subset \operatorname{Ker} f \subset H$. This means that $\operatorname{Ker} f$ is the 'largest' normal subgroup of G contained in H.*

PROOF $g \in \operatorname{Ker} f$ if and only if P_g = the identity permutation if and only if $gx_iH = x_iH$ for all i if and only if $gxH = xH$ for all

$x \in G$. Taking $x = e$, we have

$$gH = H .$$

Thus $g \in H$. Hence $\text{Ker}\, f \subset H$.

Now take N to be any normal subgroup of G with $N \subset H$. Let $a \in N$ and $g \in G$. Then $g^{-1}ag \in N \subset H$. Thus $g^{-1}ag \in H$. Hence

$$agH = gH.$$

Now g is any element of G. Thus $a \in \text{Ker}\, f$. Thus $N \subset \text{Ker}\, f$. Altogether we have $N \subset \text{Ker}\, f \subset H$. This concludes the proof.

A well known special case of the above is Cayley's theorem which we give as

THEOREM 4.6.2. *Any finite group G is isomorphic to a subgroup of the symmetric group S_n for some $n \leq |G|$.*

PROOF In the above theorems take $H = \{e\}$, the identity subgroup of G. Then we have a homomorphism $f : G \to S_n$, where $n = |G|$, and $\text{Ker}\, f \subset H = \{e\}$. Thus $\text{Ker}\, f = \{e\}$.

By exercise 4.31, f is a monomorphism. Alternatively, by theorem 4.5.3, we have

$$\text{Im}\, f \cong \frac{G}{\text{Ker}\, f} = \frac{G}{\{e\}} \cong G .$$

The theorem follows.

Note. When we have a permutation representation of the group G according to definition 4.6.1, say that given by the homomorphism $f ; G \to S_n$, it is usual to call G a **permutation group**. This is done even when f is **not** a monomorphism so that G is **not** isomorphic to a subgroup of S_n.

G is said to **act** on the set S according to:

$$g(s) = f(g)(s) ,$$

where $g \in G$ and $s \in S$.

When f is **not** a monomorphism, more than one $g \in G$ has the same action on S, i.e. $f(g_1)(s) = f(g_2)(s)$ for **all** $s \in S$, where $g_1 \neq g_2$.

For the remainder of this section we shall use the term permutation group in this sense.

The following ideas prove useful.

DEFINITION 4.6.2. *Let G be a permutation group on the set $S = \{1, 2, 3, 4, \ldots, n\}$. Let $s \in S$.*

(1) The ***orbit*** *of s under the action of G is the set*

$$G(s) = \{g(s) \mid g \in G\} .$$

The ***length*** *of the orbit is $|G(s)|$.*

(2) The ***stabiliser*** *of s is the set*

$$G_s = \{g \in G \mid g(s) = s\} .$$

(3) G is ***transitive*** *on S if S itself is the only orbit.*

THEOREM 4.6.3. *Let G be a permutation group on S.*

(1) The orbits partition S.

(2) G_s is a subgroup of G for each $s \in S$.

(3) If $t = g(s)$ for some $g \in G$, where $s, t \in S$, then

$$gG_sg^{-1} = G_t.$$

(4) $|G| = |G_s||G(s)|$.

PROOF See the solutions to exercises 4.22, 4.23, 4.24, and 4.25.

Note that in the exercises and solutions, we take the narrow definition of permutation group as a **subgroup** of S_n. However, the results and proofs remain the same for the more general definition just given.

One of the most important applications of permutation groups is to combinatorics. In this branch of mathematics the central problems are: does a particular arrangement of objects exist, and, if it does, how many essentially different such arrangements are there?

In particular it is of importance to count the number of possible physical configurations of certain objects in space, e.g. configurations of molecules. The Burnside counting theorem is often useful for this purpose.

We close this section with a proof of this result and a simple application.

THEOREM 4.6.4. (The Burnside Counting Theorem.)

Let G be a permutation group acting on the set S. Let $F(g) = \{s \in S \mid g(s) = s\}$, where $g \in G$. This is the set of fixed points of $g \in G$. Let t be the number of orbits of G on S. Then

$$t|G| = \sum_{g \in G} F(g) .$$

PROOF Let $T = \{(g, s) \in G \times S \mid g(s) = s\}$.

Here $G \times S$ is the cartesian product of the sets G and S; that is the set of all ordered pairs (g, s) with $g \in G$ and $s \in S$. Hence T is the subset of $G \times S$ consisting of all ordered pairs (g, s), where g and s are related by the fact that g leaves s fixed:

$$g(s) = s.$$

We count the number of elements in T in 2 different ways. First keep s fixed and let g vary so that $g(s) = s$, then let s run through S. This gives:

$$|T| = \sum_{s \in S} |G_s| .$$

Now we keep g fixed and let s vary so that $g(s) = s$, then let g run through G. This gives

$$|T| = \sum_{g \in G} |F(g)| .$$

Thus

$$\sum_{s \in S} |G_s| = \sum_{g \in G} |F(g)| .$$

Let $s_1, s_2, s_3, \ldots, s_t$ be representatives of the t orbits (i.e. we take just one s_i from each orbit). Then

$$\sum_{i=1}^{t} \sum_{s \in G(s_i)} |G_s| = \sum_{g \in G} |F(g)| .$$

By theorem 4.6.3 (3), if $s \in G(s_i)$, then

$$G_{s_i} = gG_sg^{-1} ,$$

where $g(s) = s_i$. Thus

$$|G_{s_i}| = |G_s| ,$$

when $s \in G(s_i)$. Hence

$$\sum_{i=1}^{t} |G(s_i)||G_{s_i}| = \sum_{g \in G} |F(g)| .$$

By theorem 4.6.3 (4),

$$|G(s_i)||G_{s_i}| = |G| .$$

Hence finally

$$t|G| = \sum_{g \in G} F(g) .$$

EXAMPLE 4.6.1. Let us consider the problem of arranging 2 black and 2 white beads on a circular wire. (In a physical application the beads might represent parts of a molecule.) Two arrangements are to be regarded as essentially the same if one can be changed into the other either by sliding beads round the wire or by moving the ring in space so that adjacent beads remain adjacent. Thus we can place the beads at the corners of a square.

There would be 4! arrangements if the beads were all distinguishable, but since they are indistinguishable in pairs we actually get 4!/2!2! = 6 arrangements as shown in Diagram 4.6.1.

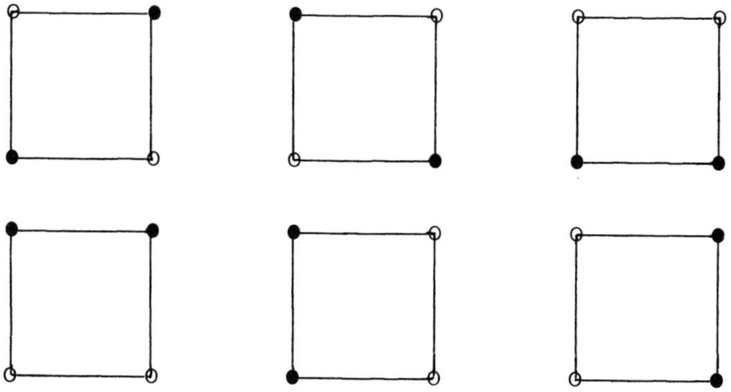

Diagram 4.6.1

However, the number of essentially distinct arrangements according to our criteria is equal to the number t of orbits of the symmetry group D_4 of the square acting as a permutation group on the set of the 6 arrangements shown in Diagram 4.6.1.

By theorem 4.6.4,

$$t = \frac{1}{|D_4|} \sum_{g \in D_4} |F(g)| \ .$$

(For the properties of the dihedral group D_4, which we quote, see later in this chapter or Chapter 7.)

We need $|F(g)|$ for each $g \in D_4$. By exercise 4.6, conjugate g have the same $|F(g)|$.

Let P be the axis perpendicular to the plane of the square. Let L be the line in the plane of the square, through its centre, and parallel to a side.

There are 5 conjugacy classes as represented by the following elements.

	Representative	**Number in the Class**
(1)	Counter-clockwise rotation through 90 degrees about axis P.	2
(2)	As above through 180 degrees.	1
(3)	Rotation through 180 degrees about axis L.	2
(4)	Rotation through 180 degrees about a diagonal.	2
(5)	The identity.	1

Now (1) fixes none of the 6 arrangements, (2) fixes 2, (3) fixes 2, (4) fixes 2, and (5) fixes all 6 arrangements. (Use Diagram 4.6.1.)

Since $|D_4| = 8$, we have:

$$t = \frac{1}{|D_4|} \sum_{g \in D_4} |F(g)| = \frac{1}{8}(2(0) + 1(2) + 2(2) + 2(2) + 1(6))$$

$$= \frac{1}{8}(0 + 2 + 4 + 4 + 6) = 16/8 = 2.$$

Hence $t = 2$. Thus there are essentially only 2 distinct ways of placing 2 black and 2 white beads at the corners of a square in space.

4.7 Subgroups of Factor Groups

If we are to exploit our knowledge of the structure of the factor group, we want to know how the subgroups of the factor group $\frac{G}{N}$ are related to the subgroups of G. This is the content of

THEOREM 4.7.1. *There is a one-one correspondence t between the subgroups of G containing the normal subgroup N and the subgroups of the factor group $\frac{G}{N}$. This enables us to write any subgroup H' of $\frac{G}{N}$ in the form $\frac{H}{N}$, where H is the subgroup corresponding to H' in the above one-one correspondence t. In fact $H = \ell^{-1}(H')$, where ℓ is the natural epimorphism $G \to \frac{G}{N}$, and $\ell^{-1}(H') = \{x \in G \mid \ell(x) \in H'\}$. We have $\ell(H) = H'$. Moreover, $H' = \frac{H}{N}$ is normal in $\frac{G}{N}$ if and only if H is normal in G.*

PROOF Let X be the set of all subgroups of G containing N. Let Y be the set of all subgroups of $\frac{G}{N}$. We define a function $t : X \to Y$ by $t(H) = \ell(H)$, and a function $s : Y \to X$ by $s(H') = \ell^{-1}(H')$. We use now the proposition from set theory that says that if we have sets X and Y with functions $t : X \to Y$ and $s : Y \to X$ such that ts is the identity function on Y and st is the identity function on X, then t and s are inverse one-one correspondences.

Now $ts(H') = \ell(\ell^{-1}(H'))$. Certainly $\ell(\ell^{-1}(H')) \subset H'$. Suppose $h' \in H'$. Then for some $g \in G$ we have $\ell(g) = h'$. Thus

$$g \in \ell^{-1}(H') .$$

Hence $\quad h' \in \ell(\ell^{-1}(H'))$.

Thus $\quad H' \subset \ell(\ell^{-1}(H'))$.

Combining this with the previous inclusion we have

$$H' = \ell(\ell^{-1}(H')) .$$

Hence $\quad ts(H') = H'$,

and so ts is the identity on Y. Now

$$st(H) = \ell^{-1}(\ell(H)) = H(\text{Ker}\,\ell)$$

by the result of exercise 4.29. Also Ker $\ell = N$, by theorem 4.5.3 (2) Thus

$$st(H) = HN.$$

Now $H \supset N$ by hypothesis. Hence

$$HN = H\ .$$

Therefore

$$st(H) = H\ .$$

It follows that st is the identity on X. Thus t defines a one-one correspondence as required in the theorem.

In the correspondence $H \leftrightarrow H'$, $H \supset N$, we have $H = \ell^{-1}(H')$. Thus H' consists of all cosets of the form hN where $h \in H$. For this reason it is customary to write H' as $\frac{H}{N}$. It follows that the subgroups of $\frac{G}{N}$ are precisely of the form $\frac{H}{N}$, where H is a subgroup of G containing N. The last part of the theorem follows from exercise 4.30 applied to $\ell : G \to \frac{G}{N}$.

The above result shows that from a knowledge of the factor grou $\frac{G}{N}$ we can at least get some knowledge of those subgroups of G that contain N.

4.8 Factor Groups of Factor Groups

Suppose we construct the factor group $\frac{G}{N}$, then take a normal subgroup $\frac{H}{N}$ of $\frac{G}{N}$, and then construct the factor group

$$\frac{\left(\frac{G}{N}\right)}{\left(\frac{H}{N}\right)}$$

of the factor group $\frac{G}{N}$.

Do we get anything essentially new?

Let us consider the natural epimorphisms ℓ_1 of G onto $\frac{G}{N}$ and ℓ_2 of $\frac{G}{N}$ onto $\frac{G}{N} / \frac{H}{N}$. The combined epimorphism $\ell_2 \circ \ell_1$ is an epimorphism from G onto $\frac{G}{N} / \frac{H}{N}$.

$$\ell_2 \circ \ell_1 : G \to \frac{G}{N} \to \frac{G}{N} / \frac{H}{N} .$$

It follows at once from theorem 4.5.3 that

$$\frac{\left(\frac{G}{N}\right)}{\left(\frac{H}{N}\right)} \cong \frac{G}{\text{Ker } \ell_2 \circ \ell_1} .$$

Thus we do not get anything new. The factor group of a factor group of G is essentially just a factor group of the original group G.

We can say a little more than this, however. The kernel of $\ell_2 \circ \ell_1$ turns out to be H. Thus we get the neat result:

THEOREM 4.8.1.

$$\frac{\left(\frac{G}{N}\right)}{\left(\frac{H}{N}\right)} \cong \frac{G}{H} .$$

PROOF To conclude the proof we have to show that Ker $\ell_2 \circ \ell_1 = H$. To see this, note that Ker $\ell_2 \circ \ell_1$ consists of all elements of G which are mapped by $\ell_2 \circ \ell_1$ onto the identity $\frac{H}{N}$ of the group

$$\frac{\left(\frac{G}{N}\right)}{\left(\frac{H}{N}\right)} .$$

Now ℓ_2 maps Ker ℓ_2 onto the identity $\frac{H}{N}$ of $\frac{\left(\frac{G}{N}\right)}{\left(\frac{H}{N}\right)}$, and by theorem 4.5.3 (2) Ker $\ell_2 = \frac{H}{N}$ as a subgroup of $\frac{G}{N}$. Thus Ker $\ell_2 \circ \ell_1$ consists of all elements in G mapped by ℓ_1 onto the subgroup $\frac{H}{N}$ of $\frac{G}{N}$. Hence

$$\text{Ker } \ell_2 \circ \ell_1 = \ell_1^{-1}\left(\frac{H}{N}\right) .$$

By theorem 4.7.1, this is precisely H.

4.9 Groups of Order p^2; p Prime

For some time now we have digressed from our central purpose of finding all groups of orders 1 to 15. The digressions have concerned themselves with structural problems, but not obviously connected with the main problem.

In this section we move closer to our central purpose by using some of our apparently unrelated results to prove the following theorem.

THEOREM 4.9.1. *All groups of order* p^2, *where* p *is a prime, are abelian.*

Besides showing that all groups of order 9 are abelian and so helping us to find the possible groups of order 9, theorem 4.9.1 also gives information about groups lying outside our range, such as those of order 25, 49, 121, 169, etc.

Let us give a proof of the theorem.

PROOF Let G be order p^2. By theorem 4.4.2 (3), $|Z(G)| = p^s > 1$. Now $|G| = p^2$. Hence by Lagrange's theorem 3.1.1, $|Z(G)|$ divides p^2.

Putting the two facts together we deduce that $|Z(G)|$ is p or p^2. If $|Z(G)| = p^2$, then $Z(G) = G$, and G is abelian by 4.4.2 (1).

Now let $|Z(G)| = p$. By theorem 4.4.2 (1), $Z(G)$ is normal so we can construct the factor group $\frac{G}{Z(G)}$ whose order is $\frac{p^2}{p} = p$. By theorem 3.1.3, $\frac{G}{Z(G)}$ is cyclic, order p, generated by $\bar{g} = gZ(G)$, say. Thus

$$\frac{G}{Z(G)} = \{\bar{g}, \bar{g}^2, \bar{g}^3, \ldots, \bar{g}^{p-1}, \bar{g}^p = \bar{e}\} .$$

Note that

$$\bar{g}^r = \overline{(g^r)} = g^r Z(G).$$

This means that G consists of the cosets

$$gZ(G), g^2 Z(G), g^3 Z(G), \ldots, g^{p-1}Z(G), g^p Z(G) = Z(G) .$$

Thus

$$G = gZ(G) \cup g^2 Z(G) \cup \ldots \cup g^{p-1}Z(G) \cup Z(G) .$$

Now let x, y be any two elements of G. Then

$$x \in g^i Z(G)$$

and

$$y \in g^j Z(G)$$

for some $1 \leq i \leq p$, $1 \leq j \leq p$. Thus

$$x = g^i c_1$$

and $$y = g^j c_2$$

for certain elements c_1 and $c_2 \in Z(G)$. This means that c_1 and c_2 commute with all elements of G. Thus

$$xy = g^i c_1 g^j c_2 = g^i g^j c_1 c_2 = g^{i+j} c_1 c_2$$

and $$yx = g^j c_2 g^i c_1 = g^j g^i c_2 c_1 = g^{i+j} c_1 c_2 .$$

Hence

$$xy = yx .$$

This means that G is abelian. Then

$$Z(G) = G ,$$

and $$|Z(G)| = p^2.$$

However, we assumed that $|Z(G)| = p$. This contradiction shows that this latter case cannot arise. Thus

$$|Z(G)| = p^2$$

and G is abelian.

Now in order to make full use of theorem 4.9.1, in particular to use it to find all possible groups of order p^2, we must know more about the structure of abelian groups. We already know the structure of a special class of abelian groups, namely the cyclic groups. What we really need is some theorem which shows how abelian groups may be built from some simple type of abelian group. The cyclic group is an obvious candidate for the 'building block'. However, at the moment, how to put together the blocks to form an arbitrary abelian group is not so clear. Nevertheless in the next chapter we shall solve this problem completely.

However, first we are going to look at some groups that illustrate many of the ideas discussed so far. These groups, the so-called finite point groups, are not only of considerable interest in themselves, but are also of great importance for scientists. The finite point groups occur in nature as the subgroups of the symmetry groups of certain physical configurations such as crystals and molecules.

4.10 Symmetry and the Orthogonal Group

For the rest of this chapter we shall be working within the usual 3-dimensional Euclidean space over the real numbers. We denote this space by $\underline{\mathbf{R}}^3$. Our axes will be mutually orthogonal straight lines with origin O, i.e. a Cartesian coordinate system. Points will be denoted by position vectors such as $\underline{v} = (x_1, x_2, x_3)$, where x_1, x_2, x_3 are the usual Cartesian coordinates of the point.

We assume that the reader is familiar with the elementary linear algebra and geometry contained in a book such as [G]. For further details of other geometrical results quoted without proof, the reader is referred to [31] and [34].

It will be helpful if the reader has available models of the regular polyhedra such as the tetrahedron, icosahedron, etc. Instructions how to make such models can be found in [32].

As indicated in Chapter 1 the symmetry of an object in space is described formally by its symmetry group. This is the group of 'movements' of the object in space which leave the object apparently fixed in space (but not only physical movements of the object as a whole, such as rotations). These 'movements' are transformations of the space, i.e. one-one, onto functions f of $\underline{\mathbf{R}}^3$ into $\underline{\mathbf{R}}^3$, which leave the **distances** between points unaltered.

Since distance or length may be described in terms of scalar product and conversely, we are interested in the group of transformations f with the property that $f(\underline{u}).f(\underline{v}) = \underline{u}.\underline{v}$.

The group of all such transformations, which **also leave the origin** O **fixed** is called the orthogonal group $O(\underline{\mathbf{R}}^3)$. The transformations are called **orthogonal** transformations.

We are interested in all the **finite** subgroups of $O(\underline{\mathbf{R}}^3)$. Such groups are called **finite point groups**. They are particularly important in science because the symmetry groups of molecules, and other such physical configurations consisting of a finite number of point objects in space, are finite point groups. (See [31] for proof.)

The object of the rest of this chapter is to classify the finite point groups and to give concrete representations of some of them as subgroups of symmetry groups of certain geometrical configurations such as the regular polyhedra.

It can be shown (see [34]) that any orthogonal transformation is a linear transformation, in the vector space sense, of $\mathbf{R}^3 \to \mathbf{R}^3$. As such it can be represented by a 3×3 matrix A over $\mathbf{R}$. Then $\underline{v} \to f(\underline{v})$ may be replaced by $\underline{v} \to A\underline{v} = A\begin{pmatrix} x_1 \\ x_2 \\ x_3 \end{pmatrix}$.

Note that A has the property that $AA' = I_3$, where A' is the transpose of A. In other words $A' = A^{-1}$. Such a matrix is called an **orthogonal** matrix. From $AA' = I_3$ we deduce that the determinant $|A| = \pm 1$.

If $|A| = +1$ the transformation is a rotation about some axis through O. This result due to Euler may be proved as follows.

THEOREM 4.10.1. *Let f be a linear transformation $\mathbf{R}^3 \to \mathbf{R}^3$ represented relative to the usual Cartesian coordinate system by the* ***orthogonal*** *matrix A. Let $|A| = +1$. Then f is a rotation about a fixed axis.*

PROOF

$$\begin{aligned} |A - I_3| &= |A' - I_3| = |A^{-1} - I_3| = |A^{-1}(I_3 - A)| \\ &= |A^{-1}||I_3 - A| = |A^{-1}||-I_3(A - I_3)| \\ &= |A'||-I_3||A - I_3| = 1.(-1)^3 . |A - I_3| \\ &= -|A - I_3| . \end{aligned}$$

Thus $2|A - I_3| = 0$. Hence

$$|A - I_3| = 0.$$

This means that 1 is an eigenvalue of A. Let $\underline{v}$ be the corresponding eigenvector. Then

$$A\underline{v} = \underline{v}$$

and $$f(\underline{v}) = \underline{v} .$$

Hence f is a transformation fixing distances and fixing the line in the direction $\underline{v}$. Thus f is a rotation about the line $\langle \underline{v} \rangle$.

If $|A| = -1$, then the transformation is an **improper** rotation or **rotation-inversion**.

If $|A| = -1$, we may write

$$A = (-I_3)(-A) = (-A)(-I_3) ,$$

where $-A$ is now a rotation and $F = -I_3$ is the so-called **inversion** in the origin: $\underline{v} \to F(\underline{v}) = -\underline{v}$.

Thus an improper rotation can always be written as a product of a rotation and the inversion F.

Note that the product of a rotation and an improper rotation is another improper rotation, and the product of 2 improper rotations is a rotation.

The set of all rotations forms a subgroup of $O(\underline{\mathbf{R}}^3)$ called the **rotation subgroup** of $O(\underline{\mathbf{R}}^3)$ or the **special orthogonal group** denoted by $SO(\underline{\mathbf{R}}^3)$.

A rotation through the angle α in the clockwise sense about the axis in the direction of the unit vector $\underline{k}$ is often denoted by $C_{\underline{k}}(\alpha)$, $0 \leqq \alpha < 2\pi$.

Since $gC_{\underline{k}}(\alpha)g^{-1} = C_{g(\underline{k})}(\alpha)$,

where $g \in SO(\underline{\mathbf{R}}^3)$ and $g(\underline{k})$ is the vector into which $\underline{k}$ is transformed by the rotation g, it follows that a class of conjugate elements in $SO(\underline{\mathbf{R}}^3)$ consists of all rotations through the same angle α.

Only rotations about the same axis commute.

A reflection in a plane perpendicular to $\underline{k}$ is usually denoted by $\sigma_{\underline{k}}$. Since

$$FC_{\underline{k}}(\pi + \beta) = FC_{\underline{k}}(\pi)C_{\underline{k}}(\beta) = \sigma_{\underline{k}}C_{\underline{k}}(\beta)$$

any improper rotation $FC_{\underline{k}}(\alpha)$ may be written: $\sigma_{\underline{k}}C_{\underline{k}}(\alpha - \pi)$. It is usual to write $S_{\underline{k}}(\alpha)$ for $\sigma_{\underline{k}}C_{\underline{k}}(\alpha)$.

Thus any improper rotation is equal to the product of a rotation about an axis $\underline{k}$ and the reflection in a plane perpendicular to $\underline{k}$.

In the orthogonal group $O(\underline{R}^3)$ we find the conjugacy classes as follows.

Let $h = Fg = gF$, where h is an improper rotation and g is a rotation. Then

$$hC_{\underline{k}}(\alpha)h^{-1} = FgC_{\underline{k}}(\alpha)g^{-1}F^{-1} = FC_{g(\underline{k})}(\alpha)F$$
$$= C_{g(\underline{k})}(\alpha)F^2 = C_{g(\underline{k})}(\alpha)$$
$$= C_{(-h(\underline{k}))}(\alpha) \; .$$

Thus $$hC_{\underline{k}}(\alpha)h^{-1} = C_{(-h(\underline{k}))}(\alpha) \text{ if } \begin{cases} h \in O(\underline{R}^3) \\ h \notin SO(\underline{R}^3). \end{cases}$$

$$gC_{\underline{k}}(\alpha)g^{-1} = C_{g(\underline{k})}(\alpha) \quad \text{if} \quad g \in SO(\underline{R}^3).$$

Similarly:

$$hS_{\underline{k}}(\alpha)h^{-1} = S_{(-h(\underline{k}))}(\alpha)$$

$$gS_{\underline{k}}(\alpha)g^{-1} = S_{g(\underline{k})}(\alpha) \; .$$

Thus the conjugacy classes consist of either all rotations through a given angle, or all improper rotations through a given angle.

4.11 Classification of the Finite Rotation Groups

Let G be a finite point group of rotations of $\underline{R}^3$. We consider the effect of G on the points on the surface of a sphere M of unit radius centred at the origin.

Let $e \neq g \in G$. Then the axis of rotation of g cuts the sphere at 2 points called **poles** with position vectors $\underline{v}$ and $-\underline{v}$. These are the only points on M left fixed by g. Let $O(\underline{v})$ denote the order of the cyclic group $\langle h \rangle$ of all rotations with the same pole $\underline{v}$. (It is fairly easy to check that this set of rotations does form a cyclic group.) Then $h, h^2, \ldots, h^{O(\underline{v})-1}$ have the same poles $\underline{v}$ and $-\underline{v}$. We say that the poles $\underline{v}$ and $-\underline{v}$ have **multiplicity** $O(\underline{v}) = O(-\underline{v})$.

Note that $\underline{v}$ is left fixed by just the $O(\underline{v})$ rotations e (the identity), h, h^2, ..., $h^{O(\underline{v})-1}$.

Let P be the set of all poles of G described by their position vectors. Then G is a permutation group on the set P. To see this, let $\underline{v}$ be the position vector of a pole of $e \neq h \in G$. Then $h(\underline{v}) = \underline{v}$.

Let $e \neq g \in G$. Then $(ghg^{-1})g(\underline{v}) = gh(\underline{v}) = g(\underline{v})$. Thus $g(\underline{v})$ is a pole of the rotation $ghg^{-1} \in G$.

Now define the function $f : G \to S_P$ (the symmetric group on the set of poles) by:

$$f(g)(\underline{v}) = g(\underline{v}) ,$$

where $g \in G$ and $\underline{v} \in P$.

It is straightforward to check that f is a homomorphism of G into S_P. Thus G is a permutation group on the set P.

As in definition 4.6.2 we let $G(\underline{v})$ be the orbit of $\underline{v}$ under the action of G, and $G_{\underline{v}}$ be the stabiliser of $\underline{v}$, where $\underline{v} \in P$.

Let W be the subset of the Cartesian product $G \times P$ defined by:

$$W = \{(g, \underline{v}) \mid e \neq g \in G,\ \underline{v} \in P,\ g(\underline{v}) = \underline{v}\} .$$

***Note** that $|G_{\underline{v}}| = O(\underline{v})$, and, by theorem 4.6.3 (4), $|G| = |G_{\underline{v}}||G(\underline{v})|$

Since each $e \neq g \in G$ has exactly 2 poles $\underline{v}$ and $-\underline{v}$,

$$|W| = 2(|G| - 1) .$$

Let $\{\underline{v}_1, \underline{v}_2, \ldots, \underline{v}_k\}$ be a set of poles one from each orbit of P under the action of G. Let $n_i = |G_{\underline{v}_i}| = O(\underline{v}_i)$ and $m_i = |G(\underline{v}_i)| =$ length of the orbit containing $\underline{v}_i$.

By theorem 4.6.3 (3) $|G_{\underline{v}}| = |G_{\underline{v}_i}|$ for all $\underline{v} \in G(\underline{v}_i)$, that is for all $\underline{v}$ in the **same** orbit as $\underline{v}_i$. Thus

$$|W| = \sum_{\underline{v} \in P} \{|G_{\underline{v}}| - 1\} = \sum_{i=1}^{k} m_i(n_i - 1)$$

$$= \sum_{i=1}^{k} (|G| - m_i),$$

since by *note above $m_i n_i = |G|$. Thus

$$2(|G| - 1) = \sum_{i=1}^{k} (|G| - m_i) .$$

Divide by $|G|$ to get:

$$2 - 2/|G| = \sum_{i=1}^{k} (1 - \frac{1}{n_i}) .$$

(Thus $\sum_{i=1}^{k} \left(1 - \frac{1}{n_i}\right) < 2$.)

We may assume $|G| > 1$. Thus

$$1 \leqq 2 - 2/|G| < 2.$$

Now each $n_i \geqq 2$; hence $1/2 \leqq 1 - \frac{1}{n_i} < 1$.

Thus k must be 2 or 3.

Let $k = 2$. Then $2 - 2/|G| = (1 - \frac{1}{n_1}) + (1 - \frac{1}{n_2})$.

Thus $2 = |G|/n_1 + |G|/n_2 = m_1 + m_2$. Hence

$$m_1 = m_2 = 1$$

and $\quad n_1 = n_2 = |G|$.

In this case G has only one axis of rotation. Hence G is a cyclic group C_n for some n.

Let $k = 3$. Take $n_1 \leqq n_2 \leqq n_3$. If $3 \leqq n_1$, then

$$\sum_{i=1}^{3} \left(1 - \frac{1}{n_i}\right) \geqq \left(1 - \frac{1}{3}\right) + \left(1 - \frac{1}{3}\right) + \left(1 - \frac{1}{3}\right) = 2 .$$

Contradiction. Thus $n_1 = 2$. Hence

$$2 - 2/|G| = \frac{1}{2} + \left(1 - \frac{1}{n_2}\right) + \left(1 - \frac{1}{n_3}\right) .$$

Thus: $\quad \frac{1}{2} + 2/|G| = \frac{1}{n_2} + \frac{1}{n_3}$.

If $4 \leqq n_2$ then $\frac{1}{n_2} + \frac{1}{n_3} \leqq \frac{1}{4} + \frac{1}{4} = \frac{1}{2}$.

Contradiction. Thus $n_2 = 2$ or 3.

Let $n_2 = 2$. Then $n_1 = n_2 = 2$. Let $n_3 = n$. Then $|G| = 2n$ We have 2 orbits each consisting of n poles each of multiplicity 2, and 1 orbit consisting of 2 poles each of multiplicty n.

These conditions determine the so-called dihedral group D_n that we shall discuss in the next section.

Let $n_2 = 3$. Then $n_1 = 2$, $n_2 = 3$, and so $n_3 \geqq 3$. Thus

$$\frac{1}{2} + 2/|G| = \frac{1}{3} + \frac{1}{n_3} .$$

Hence
$$12 = |G|\left(\frac{6}{n_3} - 1\right) .$$

Thus $n_3 \leqq 5$. The only possibilities are: $n_3 = 3$, 4, or 5. Thus we have:

(1) $n_1 = 2,\ n_2 = 3,\ n_3 = 3,\ |G| = 12$.

Then G acts as a permutation group on P so that there are 2 orbits each consisting of 4 poles each of multiplicity 3, and 1 orbit consisting of 6 poles each of multiplicity 2. By symmetry the 4 poles in the one orbit lie at the corners of a regular tetrahedron on the sphere M, and the 4 poles in the other orbit lie at their antipodes.

Thus G acts as the group of rotational symmetries of a tetrahedron, the so-called tetrahedral group, which is known to be a group of order 12.

(2) $n_1 = 2,\ n_2 = 3,\ n_3 = 4,\ |G| = 24$.

In this case there is an orbit consisting of 6 poles, each of multiplicity 4, which form the corners of a regular octahedron on M. Thus G is isomorphic to the group of rotational symmetries of the octahedron, the octahedral group of order 24.

(3) $n_1 = 2,\ n_2 = 3,\ n_3 = 5,\ |G| = 60$.

There is an orbit consisting of 12 poles, each of multiplicity 5, which form the corners of a regular icosahedron on M. In this case G is isomorphic to the group of rotational symmetries of the icosahedron, the icosahedral group of order 60.

These groups will be discussed in the next section.

4.12 Examples of Finite Rotation Groups

In this section we consider the examples of the finite rotation groups furnished by subgroups of the groups of rotational symmetries of various geometrical configurations.

Notation. The axis C is called an nth order axis of the group G if the cyclic subgroup of G consisting of all rotations about C is of order n. We denote an nth order axis by $C(n)$.

If any rotation about the axis C is conjugate in G to its inverse then the axis C is said to be 2-sided in G. Otherwise the axis is one-sided. From the equations:

$$gC_{\underline{k}}(\alpha)g^{-1} = C_{g(\underline{k})}(\alpha) \quad , \qquad g \in SO(\underline{\mathbf{R}}^3) \ ,$$

$$hC_{\underline{k}}(\alpha)h^{-1} = C_{(-h(\underline{k}))}(\alpha) \ , \quad h \in O(\underline{\mathbf{R}}^3) \ , \ h \notin SO(\underline{\mathbf{R}}^3) \ ,$$

we deduce that C is 2-sided if G contains a rotation of π about any axis perpendicular to C, or a rotation-inversion about C.

Two axes $C(n)$ and $C'(n)$ are said to be equivalent if the rotation c_n' of order n about $C'(n)$ is conjugate in G to the rotation c_n or c_n^{-1} about $C(n)$. For this to happen G must contain an element that maps $C(n)$ into $C'(n)$.

(1) The Dihedral Group D_n

This is the group of rotational symmetries of a regular n-sided prism or the group of **all** symmetries of a **plane** regular n-sided polygon.

There is one $C(n)$ axis perpendicular to the plane of the polygon through its centre, and n $C(2)$ axes perpendicular to the $C(n)$ axis. These latter are usually denoted by $u_1, u_2, u_3, \ldots, u_n$.

The angle between u_i and u_{i+1} (adjacent) is π/n.

D_n consists of the unit element e, the $(n - 1)$ rotations about the $C(n)$ axis, and a rotation of π about each u_i axis. Thus

D_n has order $2n$.

The existence of u_i makes $C(n)$ 2-sided. Hence the rotations c_n^t and c_n^{n-t} are mutually conjugate, where c_n denotes the rotation of order n about the $C(n)$ axis.

If n is even, the rotations about $C(n)$ give $1 + n/2$ conjugacy classes:

$$\{e\},\ \{c_n,\ c_n^{n-1}\},\ \{c_n^2,\ c_n^{n-2}\},\ \ldots\ ,\ \{c_n^{n/2-1},\ c_n^{n/2+1}\},\ \{c_n^{n/2}\}\ .$$

If n is odd, the same rotations give $(n + 1)/2$ classes:

$$\{e\},\ \{c_n,\ c_n^{n-1}\},\ \ldots,\ \{c_n^{(n-1)/2},\ c_n^{(n+1)/2}\}\ .$$

Note from $gC_{\underline{k}}(\alpha)g^{-1} = C_{g(\underline{k})}(\alpha)$, we deduce that rotations are conjugate in D_n if they are through the same angle and there exists in D_n a rotation which takes $\underline{k}$ into the new axis of rotation.

The axes u_1, u_3, u_5, ... are all equivalent since they map into each other under the rotations c_n^t $(t = 1, 2, 3, \ldots, n)$. The axes u_2, u_4, u_6, ... are all equivalent in the same way. Also c_n maps u_{n-1} into u_1. Thus u_{n-1} is equivalent to u_1.

Hence when n is **odd** all the u_i are equivalent. In this case the rotations about them form a single conjugacy class.

For n **even** the rotations about u_1, u_2, ..., u_n give 2 conjugacy classes:

$$\{u_1,\ u_3,\ u_5,\ \ldots,\ u_{n-1}\}$$

and

$$\{u_2,\ u_4,\ \ldots,\ u_n\}\ .$$

We note that D_n is generated by c_n and a rotation b, say, through π radians about one of the axes u_i.

$$D_n = \langle c_n,\ b \mid c_n^n = e,\ b^2 = e,\ bc_n = c_n^{-1}b \rangle\ .$$

Also $D_n = HK$, where $H = \langle c_n \rangle$ is normal in D_n of index 2 and $K = \langle b \rangle$ is a subgroup of D_n of order 2. We have

$$H \cap K = \{e\}\ .$$

Under these circumstances D_n can be constructed from its subgroups H and K by a construction called a semidirect product as follows.

DEFINITION 4.13.1. *Let H and K be groups. Let $f : K \to A(H)$ be a homomorphism of K into the automorphism group of H (see exercise 4.9). Then $H \times K$, the set of ordered pairs (h, k), $h \in H$, $k \in K$, forms a group when product is defined by:*

$$(h, k)(h', k') = (hf(k)(h'), kk') .$$

This group is called the ***semidirect product*** *of H by K.*

Of course it should be checked that $H \times K$ with the given product does form a group. In fact the identity is (e, e) where e denotes the identity in both H and K, and $(h, k)^{-1}$ is $(f(k^{-1})(h^{-1}), k^{-1})$.

When f maps the whole of K onto the identity automorphism of H, we get the **direct product** of H and K. The latter is discussed in some detail in Chapter 5 and exercises 5.

We shall not discuss the semidirect product but refer the reader to the more advanced texts on group theory in the bibliography in particular [5] where many other constructions of groups are also discussed.

Before leaving the dihedral group let us exhibit it as a semidirect product.

We take $H = \langle c_n \rangle$ and $K = \langle b \rangle$ and define the homomorphism $f : K \to A(H)$ by

$$f(b) = \theta ,$$

where θ is the automorphism of H given by

$$\theta(c_n) = c_n^{-1} .$$

We leave the reader to carry out the various verifications involved.

We remark that in constructing a group G from simpler groups H and K, it is hoped that some of the properties of G can be deduced from those of H and K and the nature of the construction.

Note that D_n for $n = 3, 4, 5, 6,$ and 7 is discussed in some detail in Chapter 7.

For the remaining examples the reader is advised to have available suitable models of the regular polyhedra. These will make it easier to check the statements made below and to further study the groups involved. Limitations of space restrict us to the following summary of the facts.

(2) **The Tetrahedral Group T**

This is the group of all the rotational symmetries of a regular tetrahedron. It is not the full symmetry group. Its order is 12. There are 4 conjugacy classes. T has 3 $C(2)$ axes and 4 $C(3)$ axes.

(3) **The Octahedral Group O**

This is the group of rotational symmetries of the octahedron or of the cube. From a cube we obtain the associated octahedron by joining the centres of adjacent faces. O has order 24. There are 5 conjugacy classes, and 6 $C(2)$, 4 $C(3)$ and 3 $C(4)$ axes.

(4) **The Icosahedral Group Y**

This is the group of rotational symmetries of the dodecahedron with 12 pentagonal faces or of the icosahedron. The latter is obtained from the former by joining the centres of adjacent faces. Y has order 60. There are 5 conjugacy classes. Y has 15 $C(2)$, 10 $C(3)$ and 6 $C(5)$ axes.

In fact Y is isomorphic to the alternating group A_5; see exercise 4.19.

All the finite point groups of the **first kind**, as the finite rotation groups are often called, occur as one or other of the groups just considered, or as subgroups of these groups.

4.13 Classification of Finite Point Groups of the Second Kind

The only finite point groups remaining to be classified are those which contain improper rotations. These groups are often called finite point groups of the **second kind**.

Let G be any such group. Let $e \neq g \in G$ and let g be an improper rotation. Let $h_1, h_2, \ldots, h_n$ be all the rotations in G, with $h_1 = e$ the identity.

The set $H = \{h_1, h_2, \ldots, h_n\}$ forms a subgroup, called the rotation subgroup, of G. We note that $h_1 g, h_2 g, \ldots, h_n g$ form a set of distinct improper rotations. Let g' be any improper rotation in G. Then $g'g^{-1}$ must be a rotation. Thus

$$g'g^{-1} = h_i$$

for some i. Hence

$$g' = h_i g \ .$$

Thus

$$G = \{h_1, h_2, \ldots, h_n, h_1 g, h_2 g, \ldots, h_n g\} \ .$$

Hence
$$|G| = 2n \ ,$$
$$[G : H] = 2 \ ,$$

and H is a normal subgroup of G.

Recall that F is the inversion defined by $F(\underline{v}) = -\underline{v}$. We have 2 cases as follows.

(1) Let $F \in G$. Then we may take $g = F$ in the above and

$$G = \{h_1, h_2, \ldots, h_n, h_1 F, h_2 F, \ldots, h_n F\} \ .$$

Since $F^2 = e$ and $h_i F = F h_i$ for all h_i, $K = \{e, F\}$ forms a subgroup of G and $G = HK$, with H and K normal in G and $H \cap K = \{e\}$. Thus G is a **direct product** of H and K.

(2) Now suppose that $F \notin G$. As before

$$G = \{h_1, h_2, \ldots, h_n, h_1 g, h_2 g, \ldots, h_n g\} \ ,$$

$$H = \{h_1, h_2, \ldots, h_n\},$$

$$|G| = 2n \ ,$$

$$[G : H] = 2 \ ,$$

and H is normal in G.

Now each improper rotation $h_i g$ can be written as $t_i F$, where t_i is a rotation. Hence

$$G = \{h_1, h_2, \ldots, h_n, t_1F, t_2F, \ldots, t_nF\} .$$

Suppose that $t_i = h_j$ for some j. Then $h_ig = h_jF$. Thus

$$F = h_j^{-1}h_ig \in G .$$

Contradiction.

Hence all the t_i are different from all the h_j. Thus

$$G^* = \{h_1, h_2, \ldots, h_n, t_1, t_2, \ldots, t_n\}$$

is a **rotation** group of order $2n$ with H a normal subgroup of index 2.

Conversely, if $G^* = \{h_1, h_2, \ldots, h_n, t_1, t_2, \ldots, t_n\}$ is a rotation group of order $2n$ with $H = \{h_1, h_2, \ldots, h_n\}$ a normal subgroup of index 2, then $G = \{h_1, h_2, \ldots, h_n, t_1F, t_2F, \ldots, t_nF\}$ will be a finite point group of the second kind.

To sum up, a finite point group G of the second kind is either a direct product of a finite rotation group H by $K = \{e, F\}$ or is obtained from a finite rotation group G^* containing a subgroup H of index 2 by putting

$$G = H \cup \{tF \mid t \in G^* - H\} .$$

There are isomorphisms between some of these groups which nevertheless turn out to be **geometrically** distinct in the sense that the inversion F of order 2 may correspond to a reflection σ_k of order 2 in the isomorphic group. However, we shall not pursue this here.

4.14 Examples of Some of the Finite Point Groups of the Second Kin

We list some of the groups that occur frequently in the literature. First we make a few remarks about the notation which is standard.

Notation. If a suffix h occurs in the symbol for a group it means that there is a horizontal plane of symmetry, i.e. a plane perpendicular to the main (single) axis $C(n)$. If a suffix v occurs it means that there is at least one vertical plane of symmetry, i.e. a plane passing through the single $C(n)$ axis.

List of Groups

(1) S_{2n} consists of powers of the rotation-inversion (improper rotation) $s_{2n} = S_{\underline{k}}(\pi/n)$. It has $2n$ elements. The even powers form a subgroup C_n.

(2) C_{nh} consists of rotations and rotation-inversions about a fixed axis through all integral multiples of $2\pi/n$. It contains $2n$ elements:

$$c_n^t,\ \sigma_{\underline{k}} c_n^t = S_{\underline{k}}\left(\frac{2\pi t}{n}\right) \quad ; \qquad t = 0, 1, 2, \ldots, (n - 1)$$

(3) C_{nv} is the symmetry group of a regular pyramid on a regular n-sided polygon as base. It has order $2n$. There are $n/2 + 3$ conjugacy classes if n is even and $(n + 3)/2$ if n is odd.

(4) D_{nh} is the symmetry group of a regular prism whose cross-section is a regular n-sided polygon. It has order $4n$. There are $n + 6$ conjugacy classes if n is even and $n + 3$ if n is odd.

(5) D_{nd} (d denotes vertical reflection planes through a single rotation-inversion axis $S(2n)$) is the symmetry group of 2 equal regular n-gonal prisms, as in (4) above, joined to each other at their bases and rotated relative to each other through π/n. D_{nd} has order $4n$ and has $n + 3$ conjugacy classes.

(6) T_d is the symmetry group of the regular tetrahedron. T_d has order 24 and has 5 conjugacy classes.

(7) T_h is obtained from the tetrahedral group T by forming the direct product of T with $K = \{e, F\}$, where F is the inversion $F(\underline{v}) = -\underline{v}$. Thus $T_h = T \times K$, with product given by:

$$(g, k)(g', k') = (gg', kk') .$$

The order of T_h is 24. There are 8 conjugacy classes.

(8) O_h is the symmetry group of the cube or of the octahedron. $|O_h|$ is 48. There are 12 conjugacy classes.

(9) Y_h is the symmetry group of the regular pentagon dodecahedron or of the icosahedron. $|Y_h|$ is 120. There are 10 conjugacy classes.

As an exercise we leave the reader to investigate how the above groups are obtained from finite rotation groups by using the 2 methods discussed in section 4.13.

EXERCISES 4

1. Find which subgroups of S_3 are normal.

2. Prove theorem 4.1.3 that any subgroup H of index 2 in a group G must be normal in G.

3. Prove theorem 4.3.3.

4. Define a relation R on a group as follows. xRy if and only if $xy^{-1} \in H$, where H is a subgroup of G.

Prove that R is an equivalence relation on G and the equivalen classes coincide with the right cosets of H in G.

5. Repeat question 4 above for T defined by xTy if and only if $x^{-1}y \in H$, and the left cosets of H in G.

6. Let $f \in S_n$ be written as a product of disjoint cycles. In this product let there be n_1 cycles of length 1, n_2 cycles of length 2, n_t cycles of length t. Then $n_1 + 2n_2 + 3n_3 + \ldots + tn_t = n$.

Call the ordered t-ple $(n_1, n_2, n_3, \ldots, n_t)$ the **cycle pattern** of f. Prove that two permutations f and $g \in S_n$ are conjugate in S_n if and only if they have the same cycle pattern.

7. Let $f \in S_n$ have cycle pattern $(n_1, n_2, \ldots, n_t)$. Prove that the number of permutations $g \in S_n$ conjugate to f in S_n is equal to:

$$n!\Big/ 1^{n_1} n_1!\ 2^{n_2} n_2!\ 3^{n_3} n_3! \ldots t^{n_t} n_t! \ .$$

8. Verify the results of questions 6 and 7 above for S_3.

9. If $f : G \to G$ is an isomorphism of a group G onto itself, we call f an **automorphism** of G. Show that the set $A(G)$ of all automorphisms of G is a group when multiplication is the usual composition of functions.

10. Let $i_x : G \to G$ be the function from the group G into itself defined by $i_x(g) = xgx^{-1}$ for all $g \in G$. Prove that i_x is an automorphism of G. (i_x is called an **inner** automorphism of G.)

11. The set of all inner automorphisms of a group G is denoted by $I(G)$. Prove that $I(G)$ is a normal subgroup of $A(G)$.

12. Let G be a group. Prove that $G/Z(G) \cong I(G)$.

13. Prove that the subgroup H of a group G is normal if and only if $i_x(H) \subset H$ for all $x \in G$.

14. Prove theorem 4.4.1.

15. Let G be **any** subgroup of S_n. Let S be the subgroup of **even** permutations in G. (See exercise 3.11.) **Note** that (1) is an even permutation. Prove that either $S = G$ or $|S| = |G|/2$. (Hint: Consider the function $f : G \to \{1, -1\} \cong C_2$.)

16. Show that the alternating group A_4 (see exercise 3.11) has no subgroup of order 6.

17. Give a counter-example to show that the converse of Lagrange's theorem 3.1.1 is false.

18. Find the conjugacy classes of S_4. Hence find all normal subgroups of S_4.

19. Assume that A_5 has 5 conjugacy classes with 1, 20, 15, 12, and 12 elements in them. Prove that A_5 is a simple group, **without** assuming theorem 4.2.2.

20. Let $G \neq \{e\}$ be a simple abelian group. Prove that G is cyclic of prime order.

21. Let G be a permutation group of degree n (that is $G \subset S_n$). Let G contain an odd permutation. Show that G is either C_2 or is **not** simple.

22. Let $G \subset S_n$ be a permutation group on the set $S = \{1, 2, 3, \ldots, n\}$. Let $s \in S$. The orbit of s under G is the set $G(s) = \{f(s) \mid f \in G\}$. $|G(s)|$ is the **length** of the orbit. The set $G_s = \{f \in G \mid f(s) = s\}$ is called the **stabiliser** of s.

Prove that two orbits are identical whenever they have an element in common. Thus S is a union of disjoint orbits.

23. Prove that the stabiliser G_s is a subgroup of G for each $s \in S$.

24. With notation as in question 22 above, let $t = f(s)$ for some $f \in G$, where $s, t \in S$. Prove that $fG_sf^{-1} = G_t$.

25. Let G be as in question 20 above. Prove that

$$|G| = |G_s||G(s)| .$$

26. Compute and identify S_3/A_3. Verify theorem 4.7.1.

27. Compute and identify S_4/K_4, where K_4 is a subgroup of S_4 isomorphic to the Klein 4-group and normal in S_4. Verify theorem 4.7.1.

28. Compute and identify A_4/K_4, where K_4 is as above. Verify theorem 4.7.1.

29. Let f be a homomorphism of the group G into the group M. Let H be a subgroup of G. Prove that

$$f^{-1}(f(H)) = H(\text{Ker } f) = (\text{Ker } f)H .$$

(Recall that $f^{-1}(S) = \{g \in G \mid f(g) \in S\}$)

30. Let f be a homomorphism of G into M. Let H be a subgroup of G. Prove that $f(H)$ is a subgroup of M, which is normal in $f(G)$ if H is normal in G. Let L be a subgroup of M. Prove that $f^{-1}(L)$ is a subgroup of G, which is normal in G if L is normal in M.

31. Let f be a homomorphism of the group G into the group M. Prove that f is one-one (a monomorphism) if and only if $\text{Ker } f = \{e\}$, the identity subgroup of G.

32. Let G be a finite group. Let the map $f : G \to S_G$ (the symmetric group on the set G) be given by $f(x) = i_x$, $x \in G$, where i_x is the inner automorphism as in question 10. Show that f is a permutation representation of G in S_G. Use the Burnside counting theorem 4.6.4 to show that the number of conjugacy classes of G is

$$\sum_{x \in G} \frac{1}{[G : N_G(x)]}$$

CHAPTER 5.
FINITELY GENERATED ABELIAN GROUPS

5.1 Introduction

In order to solve our particular problem of finding all possible abelian groups of order p^2, p prime, or even the larger problem of finding all possible abelian groups of given order n, we need to consider the structure of finite abelian groups only. However, it turns out to be not much more difficult to find the structure of a wider class of abelian groups, namely those which are generated by a finite subset of their elements. We call these finitely generated abelian groups.

One way of describing a group is to specify its generators and relations. For example, the Klein 4-group is described by $\langle a, b \mid a^2 = b^2 = e, ab = ba \rangle$. This means that K_4 consists of all finite products of a and b subject to the relations $a^2 = e$, $b^2 = e$, and $ab = ba$. We have met this idea before in sections 2.6 and 3.3. We shall describe in this way the abelian groups that we shall meet.

It is convenient to have the idea of a free group. We shall only need the special case of a free abelian group. It is also convenient to use + and talk about addition rather than multiplication when working with abelian groups. 0 is then the identity of the group rather than e.

Another concept that we need is that of a **direct product** of groups. Again we need only the direct product of abelian groups.

(The direct product of non-abelian groups is considered in the exercises.) In the case of abelian groups, since we are using + and sum we talk about **direct sum** rather than direct product.

5.2 Direct Sum

DEFINITION 5.2.1. *G is a direct sum of the groups $A_1, A_2, \ldots, A_n$ if $G = A_1 \times A_2 \times \ldots \times A_n$, the Cartesian product of the A_i, and + is defined by $(a_1, a_2, \ldots, a_n) + (b_1, b_2, \ldots, b_n) = (a_1 + b_1, a_2 + b_2, \ldots, a_n + b_n)$, for $a_i, b_i \in A_i$; $i = 1, 2, \ldots, n$. (We leave the reader to check that G is a group.)* ***Note*** *that* ***group*** *means* ***abelian group***. *Unless otherwise stated we shall adopt this convention for the rest of this section.*

The concept defined above is sometimes called the **external** direct sum of $A_1, A_2, \ldots, A_n$.

G is called an **internal** direct sum of $A_1, A_2, \ldots, A_n$ if $A_1, A_2, \ldots, A_n$ are all subgroups of G and G is isomorphic to the external direct sum of $A_1, A_2, \ldots, A_n$. We write

$$G = A_1 \oplus A_2 \oplus \ldots \oplus A_n$$

for **both** the above concepts. Any distinction will be clear from the context or will not matter.

We have the following

THEOREM 5.2.1. *$G \cong A_1 \oplus A_2 \oplus \ldots \oplus A_n$ if and only if G contains subgroups $A_i' \cong A_i$; $i = 1, 2, \ldots, n$, satisfying*

(1) $$G = A_1' + A_2' + \ldots + A_n' ;$$

that is G is generated by the A_i'; $i = 1, 2, \ldots, n$.

(2) $$A_i' \cap (A_1' + A_2' + \ldots + A_{i-1}' + A_{i+1}' + \ldots + A_n') = \{0\}.$$

for $i = 1, 2, \ldots, n$.

PROOF. For simplicity we carry through the proof for the case $n = 2$. It is routine to modify this to the general case.

Let $G \cong A_1 \oplus A_2$. To establish the 'only if' part we may take G to be $A_1 \oplus A_2$. Put

$$A_1' = \{(a_1, 0) \mid a_1 \in A_1, 0 \in A_2\}$$

$$A' = \{(0, a_2) \mid 0 \in A_1, a_2 \in A_2\}.$$

We leave the reader to check that A_1' and A_2' are subgroups of $G = A_1 \oplus A_2$.

Let (a_1, a_2) be **any** element of $A_1 \oplus A_2$ then $(a_1, a_2) = (a_1, 0) + (0, a_2)$. Thus $A_1 \oplus A_2$ is generated by A_1' and A_2'. Hence condition (1) is satisfied.

Let $x \in A_1' \cap A_2'$. Then $x \in A_1'$ means $x = (a_1, 0)$ and $x \in A_2'$ means $x = (0, a_2)$. Thus

$$(0, a_2) = (a_1, 0).$$

This means that $a_1 = a_2 = 0$. Thus

$$x = (0, 0) ,$$

the zero of $A_1 \oplus A_2$. Hence condition (2) is satisfied.

To prove the converse, suppose we are given that G contains subgroups A_1' and A_2' with $A_1' \cong A_1$ and $A_2' \cong A_2$, and that

(1) $\qquad G = A_1' + A_2'$

and

(2) $\qquad A_1' \cap A_2' = \{0\}.$

We want to prove that $G \cong A_1 \oplus A_2$. Let f_1 be an isomorphism $A_1 \to A_1'$ and let f_2 be an isomorphism $A_2 \to A_2'$. Define a function $f : A_1 \oplus A_2 \to G$ by $f((a_1, a_2)) = a_1' + a_2'$, where $f_1(a_1) = a_1'$ and $f_2(a_2) = a_2'$. By condition (1) f is onto. Also

$$\begin{aligned} f((a_1, a_2) + (b_1, b_2)) &= f((a_1 + b_1, a_2 + b_2)) \\ &= (a_1 + b_1)' + (a_2 + b_2)' \\ &= a_1' + b_1' + a_2' + b_2' \\ &= a_1' + a_2' + b_1' + b_2' \\ &= f((a_1, a_2)) + f((b_1, b_2)). \end{aligned}$$

Thus f is a homomorphism.

Finally let $f((a_1, a_2)) = 0$. Then $a_1' + a_2' = 0$. Thus

$$a_1' = -a_2' .$$

Now $A_1' \cap A_2' = \{0\}$. Hence

$$a_1' = -a_2' = 0.$$

Thus $$a_1' = a_2' = 0.$$

Now f_1 and f_2 are isomorphisms. Hence

$$a_1 = a_2 = 0.$$

Then $$(a_1, a_2) = (0, 0).$$

Thus Ker f is just the identity of $A_1 \oplus A_2$. By exercise 4.31 this means that f is one-one. Altogether f is an isomorphism. Thus $G \cong A_1 \oplus A_2$.

Note that we sometimes write the content of the previous theorem more loosely as follows:

$$G = A_1 \oplus A_2 \leftrightarrow A_1 \cap A_2 = \{0\}$$

and $$G = A_1 + A_2 .$$

We have also an equivalent alternative.

THEOREM 5.2.2. *$G = A_1 + A_2$ and $A_1 \cap A_2 = \{0\}$ if and only if every element $g \in G$ can be written* ***uniquely****, apart from the order of the terms, in the form*

$$g = a_1 + a_2 ,$$

where $a_1 \in A_1$ and $a_2 \in A_2$.

PROOF Suppose $G = A_1 + A_2$ and $A_1 \cap A_2 = \{0\}$. Then from $G = A_1 + A_2$ it follows that every $g \in G$ can be written in the form $g = a_1 + a_2$, where $a_1 \in A_1$ and $a_2 \in A_2$. Suppose also $g = b_1 + b_2$. Then $a_1 + a_2 = b_1 + b_2$ with $b_1 \in A_1$ and $b_2 \in A_2$. Thus

$$a_1 - b_1 = b_2 - a_2 .$$

Hence

$$a_1 - b_1 = b_2 - a_2 \in A_1 \cap A_2 = \{0\} .$$

Thus $$a_1 - b_1 = b_2 - a_2 = 0 .$$

Hence $a_1 = b_1$ and $a_2 = b_2$. Thus $g = a_1 + a_2$ is unique.

Now suppose the converse. Every $g \in G$ can be written uniquely in the form: $g = a_1 + a_2$; $a_1 \in A_1$ and $a_2 \in A_2$. Then $G = A_1 + A_2$ is immediate. Suppose $x \in A_1 \cap A_2$. Then $x = a_1 + 0$ for some $a_1 \in A_1$, and $x = 0 + a_2$ for some $a_2 \in A_2$. By uniqueness $a_1 = 0$ and $a_2 = 0$. Thus $x = 0$. Hence $A_1 \cap A_2 = \{0\}$. This proves the converse.

The above result extends to the general case in the obvious way.

5.3 Free Abelian Groups

With the help of the direct sum construction described above, we can define a free abelian group and we can show how every finitely generated abelian group can be constructed from cyclic groups.

First we have

DEFINITION 5.3.1. *F is said to be a free abelian group if and only if F is a direct sum of infinite cyclic groups.*

Using our theorems above this is equivalent to: F is a group with the property that every $g \in F$ can be written uniquely in the form $g = a_1 + a_2 + \ldots + a_n$, where each $a_i \in U_i$, an infinite cyclic group. We recall that $U_i \cong (\mathbf{Z}, +)$ and that $U_i = \langle u_i \rangle$ where u_i is the generator of U_i.

If F is the direct sum of n infinite cyclic groups then we say that F is free on n generators. n is called the **rank** of F. From the above every $g \in F$ can be written uniquely in the form

$$g = \alpha_1 u_1 + \alpha_2 u_2 + \ldots + \alpha_n u_n,$$

where $\alpha_i \in \mathbf{Z}$; $i = 1, 2, \ldots, n$. The uniqueness is equivalent to

$$\alpha_1 u_1 + \alpha_2 u_2 + \ldots + \alpha_n u_n = 0 \leftrightarrow \alpha_1 = \alpha_2 = \ldots = \alpha_n = 0.$$

Strictly speaking we have defined only free abelian groups of **finite** rank. However, the same definition applies for free groups of infinite rank once the direct sum of an infinite number of abelian groups has been defined. (For a given free group the rank is **unique**. Free groups are isomorphic if and only if they have equal ranks.) We shall now set up the structure theory for finitely generated abelian groups through a sequence of theorems. The first two concern free abelian groups.

THEOREM 5.3.1. *Let F be a free abelian group on n generators. Then every subgroup H is also free on m generators, where $m \leqq n$.*

PROOF Let $F = \langle u_1, u_2, \ldots, u_n \rangle$. For $n = 1$, F is an infinite cyclic group. Every subgroup is also an infinite cyclic group, hence is free on one generator. Thus $m = 1$ and $m \leqq n$. The theorem is true for $n = 1$. Assume the theorem true for all F with less than or equal to $(n - 1)$ generators. We prove the theorem then holds for n generators. An appeal to induction completes the proof.

Consider $F = \langle u_1, u_2, \ldots, u_n \rangle$ and a subgroup H. If H is generated by $u_1, u_2, \ldots, u_{n-1}$, then H is free on these $(n - 1)$ generators, since $\{u_1, u_2, \ldots, u_{n-1}\}$ is a linearly independent set of generators; that is

$$\alpha_1 u_1 + \alpha_2 u_2 + \ldots + \alpha_{n-1} u_{n-1} = 0 \leftrightarrow \alpha_1 = \alpha_2 = \ldots = \alpha_{n-1} =$$

(Compare with similar ideas in the theory of vector spaces in [G].)

Now suppose $h \in H$ has the form:

$$h = \alpha_1 u_1 + \alpha_2 u_2 + \ldots + \alpha_n u_n; \quad \alpha_i \in \underline{Z}\ ; \quad \alpha_n \neq 0.$$

Let S be the set of all α_n for $h \in H$. Let μ_n be the least such positive integer of S. Then

$$\alpha_n = b\mu_n + r\ , \qquad 0 \leqq r < \mu_n\ .$$

Now $\alpha_n - b\mu_n$ is in S. Hence

$$r = \alpha_n - b\mu_n \in S\ .$$

But μ_n is the least positive integer in S. Hence $r = 0$. Thus $\alpha_n = b\mu_n$ for some integer b.

Let $\ell = \mu_1 u_1 + \mu_2 u_2 + \ldots + \mu_n u_n \in H$. Then

$$h - b\ell = \gamma_1 u_1 + \ldots + \gamma_{n-1} u_{n-1}\ ,$$

for any $h \in H$, b depends on h.

Now $h - b\ell \in N$, where N is the subgroup of H with the property that each of its elements can be expressed in terms of $u_1, u_2, \ldots, u_{n-1}$ only. This is a subgroup of the free group on $(n - 1)$ generators and so is free by the induction hypothesis.

Thus N is free on the free generators $\ell_1, \ell_2, \ldots, \ell_{m-1}$, where $(m - 1) \leqq (n - 1)$. We have $h - b\ell \in N \subset F$, and $N = \langle \ell_1, \ell_2, \ldots, \ell_{m-1} \rangle$. Thus

$$H \subset N + \langle \ell \rangle .$$

On the other hand $N \subset H$ and $\ell \in H$. Hence

$$N + \langle \ell \rangle \subset H .$$

Thus $\qquad H = N + \langle \ell \rangle = \langle \ell_1, \ell_2, \ldots, \ell_{m-1}, \ell \rangle .$

If $\beta_1\ell_1 + \beta_2\ell_2 + \ldots + \beta_{m-1}\ell_{m-1} + \beta_m\ell = 0$, $\beta_m \neq 0$, then since $\ell_1, \ell_2, \ldots, \ell_{m-1}$ involve only $u_1, u_2, \ldots, u_{n-1}$, when we express the above in terms of the linearly independent $u_1, u_2, \ldots, u_n$ we get:

$$\beta_m\mu_n = 0$$

with $\mu_n \neq 0$. Thus $\beta_m = 0$.

Now $\ell_1, \ell_2, \ldots, \ell_{m-1}$ are linearly independent because they are free generators of N. Hence

$$\beta_1 = \beta_2 = \ldots = \beta_{m-1} = 0 .$$

Thus H is free on the generators $\ell_1, \ell_2, \ldots, \ell_{m-1}, \ell$; $m \leqq n$. The theorem follows by induction on n.

THEOREM 5.3.2. *If H is a subgroup of the free group F, then H is free by the previous theorem, and we can choose a basis (i.e. a set of free generators) $\{u_1, u_2, \ldots, u_n\}$, for F, and a basis $\{v_1, v_2, \ldots, v_m\}$, $m \leqq n$,* for *H, such that*

$$v_1 = \gamma_1 u_1$$
$$v_2 = \gamma_2 u_2$$
$$\vdots$$
$$v_m = \gamma_m u_m ,$$

where the γ_i are integers and $\gamma_1 \mid \gamma_2$, $\gamma_2 \mid \gamma_3, \ldots, \gamma_i \mid \gamma_{i+1}, \ldots, \gamma_{m-1} \mid \gamma_m$

PROOF The content of this theorem is that not only is H free, but we can arrange for the generators on which F is free to be very simply related to the generators on which H is free.

Note that we use the term **basis** as in vector space theory to denote a set of generators of a group on which the group is free, that is a linearly independent set of generators; in our case, over $\mathbf{Z}$.

Start with an arbitrary basis of F, say, $f_1, f_2, \ldots, f_n$, and an arbitrary basis of H, say, $h_1, h_2, h_3, \ldots, h_m$.

We may write

$$h_i = \sum_{j=1}^{n} \alpha_{ij} f_j, \quad \alpha_{ij} \in \mathbf{Z},$$

because $h_i \in F$, $i = 1, 2, \ldots, m$. Thus

$$A \begin{pmatrix} f_1 \\ \vdots \\ f_n \end{pmatrix} \quad \begin{pmatrix} h_1 \\ \vdots \\ h_m \end{pmatrix},$$

where A is the $m \times n$ matrix (α_{ij}), $i = 1, 2, 3, \ldots, m$; $j = 1, 2, \ldots, n$. (See [G].)

We modify A until it is of the form:

$$\begin{pmatrix} \gamma_1 & 0 & \cdots & \cdots & & 0 \\ 0 & \gamma_2 & 0 & \cdots & & 0 \\ \vdots & & & & & \vdots \\ 0 & \cdots & & \gamma_m & 0 \cdots & 0 \end{pmatrix}$$

with $\gamma_i \mid \gamma_{i+1}$.

We do this by changing the basis of F and the basis of H. Corresponding to a change of basis of each group F and H we get a corresponding change in A. We use the following changes.

(1) Interchange two of the f's, say f_r and f_s. This corresponds to interchanging two columns of A. Interchange two of the h's, say h_r and h_s. This corresponds to interchanging two rows of A.

(2) Replace f_j by $f_j + \lambda f_s$ $(s \neq j)$. Then

$$h_i = \alpha_{i1}f_1 + \alpha_{i2}f_2 + \cdots + \alpha_{ij}(f_j + \lambda f_s) + \cdots + f_s(\alpha_{is} - \lambda\alpha_{ij}) + \cdots$$

This corresponds to subtracting $\lambda \times j^{\text{th}}$ column from the sth column of A.

(3) Replace h_r by $h_r + \lambda h_s$, $(r \neq s)$. Then:

$$h_r + \lambda h_s = \sum_{j=1}^{n} \alpha_{rj} f_j + \lambda \sum_{j=1}^{n} \alpha_{sj} f_j = \sum_{j=1}^{n} (\alpha_{rj} + \lambda \alpha_{sj}) f_j .$$

Hence the rth row of A is now $\alpha_{r1} + \lambda\alpha_{s1}, \ldots, \alpha_{rn} + \lambda\alpha_{sn}$, which is obtained by adding $\lambda \times s^{\text{th}}$ row to the rth row of A.

(4) Replace h_r by $-h_r$; or replace f_j by $-f_j$. These correspond to the multiplication of a row or column of A by -1.

Note that each of (1) to (4) changes bases into bases.

Hence we can get new bases for F and H by transforming the matrix by the four operations:

(1) Interchange of rows or columns.

(2) Add integral multiple of one column to another.

(3) Add integral multiple of one row to another.

(4) Multiply row or column by -1.

To prove the theorem we must transform A into the form:

$$\begin{pmatrix} \gamma_1 & 0 & \cdots & & & 0 \\ 0 & \gamma_2 & & & & 0 \\ \vdots & & \ddots & & & \vdots \\ 0 & \cdots & 0 & \gamma_m & 0 \cdots & 0 \end{pmatrix}$$

where $\gamma_i \mid \gamma_{i+1}$, by using the operations (1), (2), (3), and (4) above. This is done as follows.

Use (1), (2), and (3) to get into position (1, 1) of A an element γ_1 such that $0 \neq |\gamma_1| \leqq |\alpha_{ij}|$ for all (i, j), where $\alpha_{ij} \neq 0$, and such that this γ_1 is the smallest element that can be obtained by use of (1), (2) and (3). With γ_1 at position (1, 1) make the remaining elements of the first row as small as possible in absolute value by using (2) and the first column. Their magnitudes become less than γ_1 and so are zero by the choice of γ_1.

Use a similar process on the first column with (3) and the first row. This does not affect the first row and gives a new first column of zeros. We then have A replaced by

$$A_1 = \left(\begin{array}{c|ccc} \gamma_1 & 0 & \cdots\cdots\cdots & 0 \\ \hline 0 & & & \\ \vdots & & A' & \\ 0 & & & \end{array}\right)$$

Suppose a_{ij} is an element of A' not divisible by γ_1. Then

$$a_{ij} = q\gamma_1 + r, \qquad r \neq 0,\ r < \gamma_1.$$

Thus $$r = a_{ij} - q\gamma_1 .$$

Add the first row to the ith, then subtract $q \times$ new i^{th} column from the jth column. Then r will appear in the matrix. However, this contradicts the choice of γ_1. Thus γ_1 divides all elements of A'. Repeat the above procedure on A'. This does not affect the first row or column of A_1, and does not affect divisibility of the elements by γ_1.

$$\left(\begin{array}{c|ccc} \gamma_2 & 0 & \cdots\cdots\cdots & 0 \\ \hline 0 & & & \\ \vdots & & A'' & \\ 0 & & & \end{array}\right)$$

where γ_2 divides all elements of A'' and $\gamma_1 \mid \gamma_2$.

Not one of the matrices A, A', A'', ... etc. can be zero otherwise at some stage we would move from a basis to another basis which included a zero element. However, any basis consists of linearly independent elements. Thus such a transition cannot occur. The theorem now follows after a finite number of steps.

5.4 Structure Theorems for Finitely Generated Abelian Groups

We are now ready to obtain the promised structure theory. First we have

THEOREM 5.4.1. *Let G be an abelian group with generators $g_1, g_2, \ldots, g_n$. Then G is the direct sum of cyclic groups $\langle h_1\rangle \oplus \langle h_2\rangle \oplus \ldots \oplus \langle h_n\rangle$, where the orders of the $\langle h_i\rangle$ are γ_i for $i = 1, 2, \ldots, m$; and infinite for $i = m+1, m+2, \ldots, n$. Moreover $\gamma_i \mid \gamma_{i+1}$.*

***Note** that **some** of $\langle h_1\rangle, \ldots, \langle h_m\rangle$ may be zero; that is the corresponding γ_i is 1, and can be omitted.*

PROOF Let $F = \langle f_1, f_2, \ldots, f_n\rangle$ be a free abelian group on n free generators $f_1, f_2, \ldots, f_n$. Define a homomorphism $t : F \to G$ by $t(f_i) = g_i$, $i = 1, 2, \ldots, n$ and in general $t(\alpha_1 f_1 + \ldots + \alpha_n f_n) = \alpha_1 g_1 + \ldots + \alpha_n g_n$. This is well-defined because $\alpha_1 f_1 + \ldots + \alpha_n f_n$ specifies a **unique** element of F. The reader is left to check that t is a homomorphism.

By theorem 4.5.3, $G \cong F/\text{Ker}\, t$. Put $H = \text{Ker}\, t$. Then

$$h \in H \leftrightarrow t(h) = 0 \quad .$$

If $h = \lambda_1 f_1 + \lambda_2 f_2 + \ldots + \lambda_n f_n$, then $h \in H$ if and only if

$$\lambda_1 t(f_1) + \lambda_2 t(f_2) + \ldots + \lambda_n t(f_n) = 0 \quad .$$

That is if and only if

$$\lambda_1 g_1 + \lambda_2 g_2 + \ldots + \lambda_n g_n = 0 \quad .$$

By theorem 5.3.2 choose bases for F and H of the form: $\{u_1, u_2, \ldots, u_n\}$ and $\{v_1, v_2, \ldots, v_m\}$ such that $v_i = \gamma_i u_i$ and $\gamma_i \mid \gamma_{i+1}$ for $i = 1, 2, 3, \ldots, m$. By the isomorphism $F/H \cong G$, there exist elements $h_1, h_2, \ldots, h_n \in G$ that correspond to $u_1, u_2, \ldots, u_n$ modulo H, i.e. h_i corresponds to the coset $u_i + H$. Also G is generated by $h_1, h_2, \ldots, h_n$. Thus if $g \in G$ we have

$$g = \mu_1 h_1 + \ldots + \mu_n h_n; \quad \mu_i \in \mathbf{Z}, \quad i = 1, 2, \ldots, n.$$

Now $$g = 0 \leftrightarrow \mu_1 u_1 + \mu_2 u_2 + \ldots + \mu_n u_n \in H \ .$$

This is so if and only if

$$\mu_1 u_1 + \ldots + \mu_n u_n = \tau_1 v_1 + \ldots + \tau_m v_m \quad ,$$

for some $\tau_i \in \mathbf{Z}$, $i = 1, 2, \ldots, m$.

Now $v_i = \gamma_i u_i$. Thus the latter holds if and only if

$$\begin{aligned} \mu_1 &= \gamma_1 \tau_1 \\ &\vdots \\ \mu_m &= \gamma_m \tau_m \end{aligned}$$

and
$$\begin{aligned} \mu_{m+1} &= 0 \\ &\vdots \\ \mu_n &= 0 \end{aligned}$$

because $\{u_1, \ldots, u_n\}$ is a basis of F. Thus

$$\mu_i u_i = \tau_i \gamma_i u_i = \tau_i v_i \in H$$

for $i = 1, 2, \ldots, m$. Hence

$$\mu_1 h_1 = \mu_2 h_2 = \ldots = \mu_m h_m = 0$$

and
$$\mu_{m+1} h_{m+1} = \ldots = \mu_n h_n = 0 \quad .$$

Thus $g = \mu_1 h_1 + \ldots + \mu_n h_n = 0 \leftrightarrow$ the individual terms are zero.*

Now this means that $G = \langle h_1 \rangle \oplus \langle h_2 \rangle \oplus \ldots \oplus \langle h_n \rangle$. To see this, first note that G is generated by $h_1, h_2, \ldots, h_n$; next an expression for $g \in G$ of the form: $g = g_1 + g_2 + \ldots + g_n$, where $g_i \in \langle h_i \rangle$, is **unique**. Since if also $g = g_1' + g_2' + \ldots + g_n'$; $g_i' \in \langle h_i \rangle$, then

$$g_i = \delta_i h_i$$

and
$$g_i' = \delta_i' h_i \quad ,$$

hence
$$0 = (g_1 - g_1') + (g_2 - g_2') + \ldots + (g_n - g_n') = \sum_{i=1}^{n} (\delta_i - \delta_i')$$

Then * above shows that

$$(\delta_i - \delta_i') h_i = 0 \ .$$

But then $g_i = g_i'$; $i = 1, 2, \ldots, n$; which shows uniqueness.

Now $\mu_i h_i = 0 \leftrightarrow \mu_i u_i \in H \leftrightarrow \mu_i = \tau_i \gamma_i$ for some $\tau_i \in \mathbf{Z}$, $i = 1, 2, \ldots, m$. This is just a special case of $g \in G$, $g = 0$ considered earlier. Thus the order of $h_i = |\gamma_i|$, if $i = 1, 2, \ldots,$ and $\mu_i h_i = 0 \leftrightarrow \mu_i = 0$; $i = m + 1, \ldots, n$. Hence

$$G = \langle h_1 \rangle \oplus \ldots \oplus \langle h_m \rangle \oplus \langle h_{m+1} \rangle \oplus \ldots \oplus \langle h_n \rangle ,$$

where order $\langle h_i \rangle = |\gamma_i|$, $\gamma_i \mid \gamma_{i+1}$; $i = 1, 2, \ldots, m$; and order $\langle h_i \rangle$ = infinite; $i = m + 1, m + 2, \ldots, n$.

The next theorem tells us a little more about the cyclic summands $\langle h_i \rangle$ of finite order occurring in the decomposition given in theorem 5.4.1. It turns out that we can decompose them even more into simpler cyclic groups. Formally we have

THEOREM 5.4.2. *If G is a cyclic group of order n, then G is a direct sum of cyclic groups each of order a power of a prime p for various primes p.*

PROOF Factorise n into two relatively prime factors n_1 and n_2. Then $n = n_1 n_2$

and $(n_1, n_2) = 1$.

Then $N_1 n_1 + N_2 n_2 = 1$ for some integers N_1 and N_2.

Consider the cyclic groups $\langle n_1 g \rangle$ and $\langle n_2 g \rangle$, where $G = \langle g \rangle$, and g is of order n. Put $G_1 = \langle n_1 g \rangle$ of order n_2 and $G_2 = \langle n_2 g \rangle$ of order n_1. Now $g = N_1(n_1 g) + N_2(n_2 g)$. Thus

$$G = G_1 + G_2 .$$

Let $x \in G_1 \cap G_2$. Since $x \in G_1$, we have $n_2 x = 0$ and since $x \in G_2$ we have $n_1 x = 0$. Thus

$$x = N_1 n_1 x + N_2 n_2 x = 0 .$$

Hence $G_1 \cap G_2 = \{0\}$.

From theorem 5.2.1,

$$G = G_1 \oplus G_2.$$

Repeating this process on G_1, G_2, and so on, we get eventually

$$\langle g \rangle = \langle g_1 \rangle \oplus \langle g_2 \rangle \oplus \ldots \oplus \langle g_r \rangle ,$$

where $\langle g_i \rangle$ is of order $p_i^{t_i}$ and $n = p_1^{t_1} p_2^{t_2} \ldots p_r^{t_r}$; p_i prime. The theorem follows.

Putting theorems 5.4.1 and 5.4.2 together we conclude that any finitely generated abelian group is a direct sum of a finite number of cyclic groups each of order a power of a prime for various primes, or of infinite order.

A natural question arises now. Can we decompose our cyclic summands any further? The following theorem settles the matter.

THEOREM 5.4.3.

(1) An infinite cyclic group is ***indecomposable****; that is it* ***cannot*** *be written as a direct sum of two subgroups neither of which is the zero subgroup and neither of which is the whole group.*

(2) A finite cyclic group of order p^r*, where* p *is a prime and* r *is a positive integer, is indecomposable.*

PROOF

(1) Let $\langle g \rangle$ be an infinite cyclic group. Then any non-zero subgroup is of form

$$A = \{0, \pm mg, \pm 2mg, \pm 3mg, \ldots\}$$

and any other non-zero subgroup is

$$B = \{0, \pm tg, \pm 2tg, \pm 3tg, \ldots\},$$

where $m \neq 0 \neq t$. Thus $mtg \neq 0$ lies in both subgroups. Hence $A \cap B \neq \{0\}$ for **any** two non-zero subgroups A and B. Thus

$$\langle g \rangle \neq A \oplus B .$$

(2) Let $\langle g \rangle$ be a cyclic group of order p^r. Then the subgroups are: $\langle g \rangle, \langle pg \rangle, \langle p^2 g \rangle, \ldots, \langle p^{r-1} g \rangle$. Moreover $\langle g \rangle \supset \langle pg \rangle \supset \langle p^2 g \rangle \supset \cdots \supset \langle p^{r-1} g \rangle$. Hence if A and B are subgroups $\neq \{0\}$ or $\langle g \rangle$, then

$$A \cap B \neq \{0\} .$$

Thus $\langle g \rangle \neq A \oplus B .$

EXAMPLE 5.4.1 Let G be the abelian group described by

$$\langle x, y, z \mid x + y = y + x,\ x + z = z + x,\ y + z = z + y,\ x + y + 4z = x + 4y + z = 4x + y + z = 0 \rangle .$$

Then $G \cong F/H$,

where F is the free abelian group on the free generators x, y, z and H is the subgroup of F generated by:

$$u = x + y + 4z,$$

$$v = x + 4y + z,$$

$$w = 4x + y + z.$$

We note that a direct check shows that u, v, w are linearly independent over $\mathbb{Z}$. If not, this would emerge anyway by the appearance of a zero γ in the reduction procedure for the matrix A.

We have

$$A \begin{pmatrix} x \\ y \\ z \end{pmatrix} = \begin{pmatrix} u \\ v \\ w \end{pmatrix} ,$$

where

$$A = \begin{pmatrix} 1 & 1 & 4 \\ 1 & 4 & 1 \\ 4 & 1 & 1 \end{pmatrix} .$$

By using the row and column operations given in the text we get the following sequence of matrices:

$$A = \begin{pmatrix} 1 & 1 & 4 \\ 1 & 4 & 1 \\ 4 & 1 & 1 \end{pmatrix} \to \begin{pmatrix} 1 & 0 & 0 \\ 1 & 3 & -3 \\ 1 & -3 & -15 \end{pmatrix} \to \begin{pmatrix} 1 & 0 & 0 \\ 0 & 3 & -3 \\ 0 & -3 & -15 \end{pmatrix} \to \begin{pmatrix} 1 & 0 & 0 \\ 0 & 3 & 0 \\ 0 & -3 & -18 \end{pmatrix} \to$$

$$\to \begin{pmatrix} 1 & 0 & 0 \\ 0 & 3 & 0 \\ 0 & 0 & -18 \end{pmatrix} \to \begin{pmatrix} 1 & 0 & 0 \\ 0 & 3 & 0 \\ 0 & 0 & 18 \end{pmatrix} = \begin{pmatrix} \gamma_1 & 0 & 0 \\ 0 & \gamma_2 & 0 \\ 0 & 0 & \gamma_3 \end{pmatrix} .$$

From this we deduce that $G \cong C_1 \oplus C_3 \oplus C_{18}$. Since $18 = 2 \times 3^2$ we can decompose C_{18} to get $C_2 \oplus C_9$. Moreover, $C_1 = \{0\}$ and can be discarded.

Finally $G \cong C_3 \oplus C_2 \oplus C_9$. The order of G is $3 \times 2 \times 9 = 54$.

<u>Note</u> that C_n is in additive notation:

$$C_n = \langle a \rangle = \{0, a, 2a, \ldots, (n-1)a\} .$$

5.5 Uniqueness

Along with any structure theory such as that given in the foregoing sections there is a corresponding uniqueness question, namely if the given group is decomposed according to the structure theory do we always get the same decomposition? If not, how are the different decompositions related?

In the case of finitely generated groups these questions are answered by the following theorem which we give without proof:

THEOREM 5.5.1 *If the abelian group G is expressed as a direct sum of a finite number of cyclic groups each of order $p_i^{t_i}$, for various primes p_i, or of infinite order, then any other decomposition into indecomposable groups must involve the same number of groups of order $p_i^{t_i}$ and the same number of infinite cyclic groups.*

Thus if $G = C_\infty \oplus C_\infty \oplus C_{p_1^2} \oplus C_{p_1} \oplus C_{p_2^3}$ then any other decomposition of G into **indecomposable** summands must have the form:

$$G = C'_\infty \oplus C'_\infty \oplus C'_{p_1^2} \oplus C'_{p_1} \oplus C'_{p_2^3}$$

Note that, for example, $C_{p_1^2}$ is not necessarily the same subgroup of G as $C'_{p_1^2}$, but, as cyclic groups of the same order p_1^2, they are isomorphic.

5.6 Possible Groups of Order p^2

We know already by theorem 4.9.1 that all groups of order p^2 are abelian. Armed with the above results we can deduce now that the only groups of order p^2 are $G = C_p \oplus C_p$ and $G = C_{p^2}$. Reverting to multiplicative notation these are

$$G = \langle a, b \mid a^p = b^p = e,\ ab = ba \rangle = \langle a \rangle \langle b \rangle$$

and $$G = \langle a \mid a^{p^2} = e \rangle .$$

By the uniqueness theorem these are distinct. However, notice that the elementary consideration of the orders of certain elements suffices to show that these two groups of order p^2 are not isomorphic. In $C_p \oplus C_p$ there is no element of order p^2, whereas C_{p^2}, of course, is generated by an element of order p^2.

Considerations like this will be sufficient for our purpose. Hence we shall not require the power of the uniqueness theorem.

The theorems just obtained not only solve the problem of the groups of order p^2 they enable us to find all possible abelian groups of any order. Thus in order to find all possible groups of a given order we have to study now the **non**-abelian groups. These are more intractable. However, help is at hand in the form of some remarkable theorems discovered by Sylow in 1872. These theorems form the subject matter of the next chapter.

EXERCISES 5

1. Let $G = A \oplus B$. Prove that $G/A \cong B$ and $G/B \cong A$.

2. Let $G_1, G_2, \ldots, G_n$ be any n groups. Define a product on $G_1 \times G_2 \times \ldots \times G_n$ (the Cartesian product) by:

$(g_1, g_2, g_3, \ldots, g_n).(h_1, h_2, \ldots, h_n) = (g_1h_1, g_2h_2, \ldots, g_nh_n)$.

Show G is a group with this multiplication.

Note. G is called the **external** direct product of $G_1, G_2, G_3, \ldots, G_n$.

3. Prove that a group G is isomorphic to the external direct product $G_1 \times G_2 \times \ldots \times G_n$ if and only if

   (a) G contains subgroups $H_i \cong G_i$; $i = 1, 2, 3, \ldots, n$;

   (b) each H_i is normal in G;

   (c) $G = H_1H_2H_3 \ldots H_n = \{h_1h_2 \ldots h_n \mid h_i \in H_i\}$;

   (d) $H_i \cap H_1H_2 \ldots H_{i-1}H_{i+1} \ldots H_n = \{e\}$; the identity subgroup of G.

Note. Such a group is called an **internal** direct product of its subgroups H_i; $i = 1, 2, 3, \ldots, n$.

4. Prove that equivalent conditions to (b), (c), (d) above are:

   (a) $h_i h_j = h_j h_i$ for any $h_i \in H_i$ and $h_j \in H_j$, $i \neq j$;

   (b) every element $g \in G$ can be written **uniquely**, apart from order of the factors, in the form $g = h_1 h_2 h_3 \ldots h_n$, where $h_i \in H_i$.

Note. These results reduce to those given in the text in the case of abelian groups.

5. Let F be a free abelian group with basis u_1, u_2, u_3. Let H be the subgroup of F with generators $u_1 + u_2 + 10u_3$, $u_1 + 10u_2 + u_3$, $10u_1 + u_2 + u_3$. Find a decomposition of the group F/H as a direct sum of indecomposable cyclic groups. Find the order of F/H.

6. Let G be an abelian group generated by a, b, c, d subject to the relations:

$11a - 10b - 4c - 7d = 0$, $3a - 4b + 2c + 5d = 0$, $3a - 4b - 4c - 7d =$

Find a decomposition of G into indecomposable cyclic groups.

CHAPTER 6.
THE SYLOW THEOREMS

6.1 Introduction

We would like to know as much as possible about the subgroups of an arbitrary finite group. As might be expected very little can be said at this level of generality. However, around 1872 the Norwegian mathematician Sylow discovered some remarkable theorems about certain subgroups of any finite group. (See 'Théorèmes sur les groupes de substitutions' - Mathematische Annalen. V (1872), pp. 584 *et seq.* - Sylow.)

These theorems have been the model for many subsequent investigations of the sbugroup structure of groups. In particular they have been extended to certain infinite groups. Later we shall make use of the Sylow theorems in order to complete our search for groups of order n, for $n \leq 15$.

Before stating and proving the Sylow theorems we require some preliminary results; some of rather a technical nature.

6.2 Double Cosets

Given a group G with subgroups A and B we define a relation D on G relative to A and B by:

$$xDy \leftrightarrow x = ayb$$

for **some** $a \in A$ and **some** $b \in B$. We leave the reader to check that D is an equivalence relation on G. (Compare with the results of exercises 4.4 and 4.5 for single cosets.) The equivalence classes may be written AyB, where y is **any** element in the given class. For finite G we have:

$$G = Ax_1B \cup Ax_2B \cup \dots \cup Ax_rB ,$$

where the unions are disjoint. **For this chapter we assume all groups finite unless otherwise stated.**

Let class Ax_iB contain ℓ elements. Then $(x_i^{-1}Ax_i)B$ also contains ℓ elements. Now order $(x_i^{-1}Ax_i)$ = order A by theorem 4.3.3. By theorem 3.4.2 (3),

$$\ell = (\text{order } A)(\text{order } B)/\text{order}(x_i^{-1}Ax_i \cap B).$$

The classes AyB are called **double cosets** of G relative to A and B. AyB consists of complete **left** cosets of B in G and also of complete **right** cosets of A in G. It turns out to be useful to have an expression for the number of left cosets of B in AxB. This is the content of:

THEOREM 6.2.1. *The number of left cosets of B in AxB is the index of $x^{-1}Ax \cap B$ in $x^{-1}Ax$; denoted as usual by $[x^{-1}Ax : x^{-1}Ax \cap B]$.*

PROOF Put the elements of AxB in one-one correspondence with the elements of $x^{-1}AxB$ by the function $f\colon AxB \to x^{-1}AxB$ defined by $f(axb) = x^{-1}axb$; $a \in A$, $b \in B$. This also gives a one-one correspondence between the left cosets axB of B contained in AxB and the left cosets $x^{-1}axB$ of B contained in $x^{-1}AxB$.

Left coset decomposition of $x^{-1}Ax$ relative to its subgroup $B \cap x^{-1}Ax$ gives:

$$x^{-1}Ax = u_1(B \cap x^{-1}Ax) \cup u_2(B \cap x^{-1}Ax) \cup \dots \cup u_r(B \cap x^{-1}Ax)$$

where $r = [x^{-1}Ax : x^{-1}Ax \cap B]$ and unions are disjoint. Now u_1B, u_2B, ..., u_rB are left cosets of B contained in $x^{-1}AxB$, since $u_i \in x^{-1}Ax$ for all i. Suppose $u_iB = u_jB$ with $i \neq j$. Then

$$u_i^{-1}u_j \in B .$$

Now $u_i^{-1}u_j \in x^{-1}Ax$. Thus

$$u_i^{-1}u_j \in x^{-1}Ax \cap B .$$

Hence $\quad u_i(x^{-1}Ax \cap B) = u_j(x^{-1}Ax \cap B)$.

This contradicts *. Thus u_1B, u_2B, ..., u_rB are distinct.

Now every left coset of B contained in $x^{-1}AxB$ has the form uB, where $u \in x^{-1}Ax$. Thus $u = u_i w$, where $w \in x^{-1}Ax \cap B$, by *. Hence

$$uB = u_i wB = u_i B \; .$$

Thus $u_1 B$, $u_2 B$, ..., $u_r B$ are **precisely** the left cosets of B in $x^{-1}AxB$. Hence this number is r as required, where $r = [x^{-1}Ax : x^{-1}Ax \cap B]$.

6.3 The Sylow Theorems

First a preliminary result:

THEOREM 6.3.1. *If the order of a group G is divisible by a prime p, then G contains an element of order p, and hence G has a cyclic subgroup of order p.*

PROOF Although we introduce this result in order to start off an induction proof of the first Sylow theorem, it does have intrinsic interest and has had many proofs. Usually these involve several steps. Our combinatorial proof gives the result in one step.

Let G be of order n. We have $p \mid n$, p prime. Consider all solutions of $x_1 x_2 x_3 \ldots x_p = e$, where $x_i \in G$, and as usual e is the identity of G. Two solutions are to be considered the same if and only if $x_1 = x_1'$, $x_2 = x_2'$, ..., $x_p = x_p'$, where $x_1 x_2 \ldots x_p = e$ and $x_1' x_2' \ldots x_p' = e$. The total number of distinct solutions is then n^{p-1}, obtained by choosing the first x_1, x_2, ..., x_{p-1} arbitrarily from the n elements of G and then taking $x_p = (x_1 x_2 x_3 \ldots x_{p-1})^{-1}$.

Now consider cyclic permutations of a solution $x_1 x_2 \ldots x_p = e$. We have

$$x_p x_1 \ldots x_{p-1} = x_p (x_1 x_2 \ldots x_p) x_p^{-1} = x_p e x_p^{-1} = e \; .$$

Hence $x_p x_1 x_2 \ldots x_{p-1}$ is also a solution.

Define an equivalence relation on the set of all solutions by calling two solutions **equivalent** if and only if one can be obtained from the other by repeated cyclic permutations.

Consider one such class. Since p is a prime we obtain p distinct solutions in the class **provided not all the x_i are equal.** If the number of classes with p distinct solutions is r, and the number of classes with one solution is s, we have $s \neq 0$, since $x_1 = x_2 = \ldots = x_p = e$ is such a solution. Now $pr + s = n^{p-1}$, and $p \mid n$. Thus $p \mid (pr + s)$. Hence $p \mid s$. Now $s \neq 0$. Thus $s > 1$.

Hence there exists at least one solution

$$x_1 = x_2 = \ldots = x_p = x \neq e.$$

But then $x^p = e$, $x \neq e$. The theorem follows.

We are now in a position to prove the Sylow theorems. We have called them the first, second, and third Sylow theorems. However, this order is by no means universally agreed. It varies from author to author.

THEOREM 6.3.2. (The First Sylow Theorem) *Let G be a group of order $n = p^m s$, where $(p, s) = 1$, and p is a prime. Then G contains subgroups of orders p^i; $i = 1, 2, 3, 4, 5, \ldots, m$; and each subgroup of order p^i, $i = 1, 2, 3, \ldots, (m - 1)$, is a normal subgroup of at least one subgroup of order p^{i+1}.*

PROOF If $i = 1$, then by the previous theorem, there exists a subgroup of order p^i. To prove the theorem all we need to do is to show that if P is a subgroup of order p^i, $1 \leq i \leq (m - 1)$, then there exists a subgroup J of order p^{i+1} such that P is normal in J. Induction on i completes the proof.

Suppose then that P is a subgroup of order p^i. Now if J exists it must be contained in the normaliser K, say, of P in G. Consider K/P. If we can show that $p \mid |K/P|$, then K/P will contain an element of order p and hence a subgroup of order p by theorem 6.3.1. This subgroup of K/P will be cyclic and will have the form J/P, by theorem 4.7.1. Since J/P has order $p = |J| / |P|$ it follows that J has order $p \times p^i = p^{i+1}$. Also P is normal in J. This gives the required result. The theorem follows then if we can show that $p \mid |K/P|$.

Write G in terms of double cosets relative to P and P. We have:

$$G = PP \cup Px_2P \cup \dots \cup Px_rP ,$$

where $x_1 = e$ has been omitted. Let there be a_j left cosets of P in Px_jP. We have

$$a_j = [x_j^{-1}Px_j : x_j^{-1}Px_j \cap P]$$

by theorem 6.2.1. The order of

$$x_j^{-1}Px_j = \text{order } P = p^i .$$

Hence $a_j = 1$ or p^t, $t \geq 1$. Now $a_1 = 1$ for the double coset $PP = P$. Also $[G : P] = a_1 + a_2 + \dots + a_r$. Now $p \mid [G : P]$, since $[G : P] = p^{m-i}s$, and $i \leq (m - 1)$. Thus the number of a_j equal to 1 must be a non-zero multiple of p. Now $a_j = 1$ if and only if

$$x_j^{-1}Px_j = x_j^{-1}Px_j \cap P \leftrightarrow x_j^{-1}Px_j \subset P \leftrightarrow x_j^{-1}Px_j = P$$

(since $|x_j^{-1}Px_j| = |P|$ by theorem 4.3.3)

$$\leftrightarrow Px_j = x_jP \leftrightarrow x_j \in K = N_G(P).$$

Suppose then that for $x_j, x_i \in K$ we have $Px_jP = Px_iP$. This occurs if and only if $Px_j = Px_i$, since

$$x_jP = Px_j$$

and $\qquad x_iP = Px_i$

as $x_j, x_i \in K$.

Thus the number of **distinct** double cosets Px_jP with $x_j \in K$ is equal to the index of P in K. Thus the number of $a_j = 1$ equals $[K : P]$. Hence

$$p \mid [K : P] .$$

Now $\qquad [K : P] = |K/P|$. Thus

$$p \mid |K/P|$$

and the theorem follows.

DEFINITION 6.3.1. *The subgroups of maximal order in the above theorem, namely those of order p^m, where $|G| = p^ms$, and $(p, s) = 1$, are called the* ***Sylow p-subgroups*** *of G.*

THEOREM 6.3.3. (The Second Sylow Theorem) *In a finite group G the Sylow p-subgroups form a complete conjugate set of subgroups of G.*

PROOF Let P be a Sylow p-subgroup of G. Then $x^{-1}Px$ is also of order p^m by theorem 4.3.3. Hence $x^{-1}Px$ is also a Sylow p-subgroup of G. Thus any subgroup which is conjugate to a Sylow p-subgroup is also a Sylow p-subgroup.

Now suppose that P and P' are both Sylow p-subgroups of G. A double coset decomposition of G relative to P and P' gives:

$$G = Px_1P' \cup Px_2P' \cup \dots \cup Px_rP' \quad .$$

Let there be b_i left cosets of P' in Px_iP'. Then

$$b_i = [x_i^{-1}Px_i : x_i^{-1}Px_i \cap P']$$

by theorem 6.2.1. Now b_i divides $|x_i^{-1}Px_i| = p^m$ and so b_i is 1 or p^t, $t \geq 1$.

Now $[G : P'] = b_1 + b_2 + \dots + b_r$, and $p \nmid s = [G : P']$. Thus at least one of b_i must be 1. Suppose $b_i = 1$. Then

$$x_i^{-1}Px_i = x_i^{-1}Px_i \cap P' \; .$$

Hence $$x_i^{-1}Px_i \subset P' \; .$$

However both groups have order p^m. Thus

$$x_i^{-1}Px_i = P' \; .$$

Hence P and P' are conjugate. The theorem follows.

We have a very useful corollary of theorem 6.3.3.

THEOREM 6.3.4. *There is one unique Sylow p-subgroup of the finite group G if and only if it is normal.*

PROOF By theorem 6.3.3 P is the **only** Sylow p-subgroup of G if and only if it is self-conjugate, i.e.

$$g^{-1}Pg = P$$

for all g in G. This is true if and only if P is normal.

Note. For the following theorem we use the result that if P is a Sylow p-subgroup of G and if $P \subset H \subset G$, where H is a subgroup of G, then P is also a Sylow p-subgroup of H. This follows at once from $|P|$ divides $|H|$ and $|H|$ divides $|G|$.

THEOREM 6.3.5 (The Third Sylow Theorem) *Let G be as in the other Sylow theorems. Then the number of Sylow p-subgroups is of the form $1 + kp$, where k is a non-negative integer, and $1 + kp$ divides the order of G.*

PROOF If there is only one Sylow p-subgroup the result is trivial. Suppose there are at least two. Take S_0 as one and let $S_1, S_2, \ldots, S_r$ be the others. Transform each of them with elements of S_0. This divides them into classes, each class consisting of those S_i conjugate under transformation by elements of S_0, i.e. S_i and S_j are in the same class $\leftrightarrow S_i = x^{-1}S_jx$, where $x \in S_0$. We say that $x^{-1}S_jx$ is the **transform** of S_j by the element x of S_0. Sylow's second theorem shows that S_i is transformed into S_j, where S_i and S_j belong to the set $\{S_1, S_2, \ldots, S_r\}$. (We cannot have $x^{-1}S_ix = S_0$ with $x \in S_0$ otherwise $S_i = xS_0x^{-1} = S_0$.)

Now S_i is a Sylow p-subgroup of its normaliser K_i and, since S_i is normal in K_i, S_i is the only Sylow p-subgroup of K_i. These statements follow from theorem 6.3.4 and the note following. Thus $K_i \cap S_0 \neq S_0$ if $i \neq 0$, otherwise $S_0 \subset K_i$ and S_0 would be a second Sylow p-subgroup of K_i. Thus $K_i \cap S_0$ is a proper subgroup of S_0. Now the number of conjugates of S_i under transformation by elements of S_0 is $[S_0 : K_i \cap S_0]$ by a slight modification of the argument which established theorem 4.3.2. This number is 1 or p^{t_i}. Now $S_0 \neq K_i \cap S_0$. Hence this number is p^{t_i}, $t_i \geqq 1$. Thus if $S_1, S_2, \ldots, S_r$ split into m classes we have

$$r = p^{t_1} + p^{t_2} + \ldots + p^{t_m}\ ; \qquad t_i \geqq 1; \quad i = 1, 2, \ldots, m.$$

Thus $\quad r = kp\ , \quad k \geqq 1.$

Hence the total number of Sylow p-subgroups is 1 or $1 + kp$, $k \geqq 1$. This is $1 + kp$ with k a non-negative integer.

Since the Sylow p-subgroups form a conjugate set the number of them is the index of the normaliser of any one of them in G.

Hence this number divides the order of G.

This completes the proof of the theorem.

6.4 Applications of the Sylow Theorems

We end this chapter by giving 2 typical applications of the Sylow theorems. For ease of future reference we call the first application

THEOREM 6.4.1. *Let G be a group of order pq, where p and q are* ***distinct*** *primes with $p < q$. Then G has just one subgroup of order q. This subgroup of order q is normal in G. If p does* **not** *divide $(q - 1)$, then G is cyclic.*

PROOF There are $1 + k_1p$ Sylow p-subgroups, where $(1 + k_1p) \mid pq$. There are $1 + k_2q$ Sylow q-subgroups, where $(1 + k_2q) \mid pq$.

Now $q \nmid (1 + k_2q)$, hence $(1 + k_2q) \mid p$. However, $q > p$. Thus

$$1 + k_2q = 1 .$$

Hence there is just one Sylow q-subgroup H. By theorem 6.3.4 this subgroup H is normal.

Now suppose $p \nmid (q - 1)$. That is $q \neq 1 + k_1p$. Now $(1 + k_1p) \mid pq$ and $p \nmid (1 + k_1p)$ imply $(1 + k_1p) \mid q$. Hence

$$(1 + k_1p) = q$$

or

$$(1 + k_1p) = 1 ,$$

since q is prime. However, $(1 + k_1p) \neq q$ as above. Hence

$$1 + k_1p = 1.$$

Thus there is just one Sylow p-subgroup K, which must be normal by theorem 6.3.4.

We now use a typical argument which will recur many times in Chapter 7 and in the exercises.

$H \cap K$ is a subgroup of both H and K. Thus $|H \cap K|$ must divide both $|H| = q$ and $|K| = p$. Now $(p, q) = 1$ because p and q are distinct primes. Thus $|H \cap K|$ must be 1. Hence $H \cap K$ is $\{e\}$.

At this stage we could argue that HK is a direct product and $G = HK$. Then using this fact it is fairly easy to show that G is cyclic. However, we shall avoid the concept of direct product and argue as follows.

Let $H = \langle h \rangle$ and $K = \langle k \rangle$. Then

$$[h, k] = h^{-1}(k^{-1}hk) \in H,$$

since H is normal. But also

$$[h, k] = (h^{-1}k^{-1}h)k \in K,$$

since K is normal. Thus

$$[h, k] = h^{-1}k^{-1}hk \in H \cap K = \{e\}.$$

Hence $\quad hk = kh.$

By theorem 2.8.1 (5),

$$O(hk) = pq.$$

Thus $\quad G = \langle (hk) \rangle$ is cyclic.

This completes the proof of the theorem.

We call our second application

THEOREM 6.4.2. *Let G be any group of order 30. Then G has at least one normal subgroup.*

PROOF By Sylow's third theorem, since $30 = 2 \times 3 \times 5$, G must have
$1 + 2k_1$ subgroups of order 2
$1 + 3k_2$ subgroups of order 3
$1 + 5k_3$ subgroups of order 5 ,

where k_1, k_2 and k_3 are certain non-negative integers. Moreover, each of $(1 + 2k_1)$, $(1 + 3k_2)$, $(1 + 5k_3)$ must divide 30. By giving k_1, k_2, k_3 the values 0, 1, 2, 3, 4, 5, ... in turn we deduce that the numbers of possible subgroups of G of orders 2, 3, and 5 are:

1, 3, 5, or 15 subgroups of order 2,
1, or 10 subgroups of order 3,
and 1, or 6 subgroups of order 5.

Now suppose that there are at least 3 subgroups of order 2, 10 subgroups of order 3, and 6 subgroups of order 5.

Let H and K be any two of these **distinct** subgroups, where $|H| = p$, $|K| = q$, $p \in \{2, 3, 5\}$, $q \in \{2, 3, 5\}$ and p may be equal to q. Consider $H \cap K$. We argue exactly as in theorem 6.4.1 in order to show that $H \cap K = \{e\}$.

Thus only the identity e is common to any pair of subgroups. We count elements. The number of elements $\neq e$ contained in the 3 subgroups of order 2 is 3, in the 10 subgroups of order 3 is 20, and in the 6 subgroups of order 5 is 24. Thus the total number of elements in G must be at least $3 + 20 + 24 + 1 = 48$.

However, this contradicts $|G| = 30$. Hence there must be just one subgroup of order 2, 3, or 5. Theorem 6.3.4 assures us that this subgroup must be normal in G. Hence any group of order 30 must have at least one normal subgroup of order 2, 3, or 5.

This completes the proof of the theorem.

Note. It is particularly important to find normal subgroups in a group whose structure we wish to study. With the help of a normal subgroup we can construct a factor group and from this, assuming its structure known, we can hope to discover more about our original group by using results like theorem 4.7.1 or like those in Chapter 7, to which we now turn.

EXERCISES 6

1. Prove that no group of order 20 is simple.

2. Prove that no group of order 36 is simple.

3. Prove that no group of order 48 is simple.
(Hint. Consider subgroups of order 16).

4. Prove that no group of order 200 is simple.

5. Prove that any group of order 255 is cyclic.

6. Find all Sylow p-subgroups of A_4 for $p = 2$ and $p = 3$.

7. Let G be a finite group with just one Sylow p-subgroup for each prime p dividing $|G|$. Prove that G is the direct product of its Sylow p-subgroups. Such a group is called a finite nilpotent group. (See exercises 5.3 and 5.4.)

8. Let the order of G be $2p$, where p is an odd prime. Prove that G has precisely one subgroup of order p and **either** p subgroups of order 2, **or** one subgroup of order 2. In the latter case show that G is cyclic.

9. Let G be a finite group. Prove that G is a p-group (i.e. has order p^r for p prime and r a positive integer) if and only if every element of G has order a power of p.

10. Let G be a non-abelian group of order p^3, p a prime. Prove that $Z(G)$ has order p. Deduce that $G/Z(G)$ is abelian.

11. Let P be a Sylow p-subgroup of the finite group G. Let $N = N_G(P)$. Let U be a subgroup of G such that $N \subset U \subset G$. Prove that $N_G(U) = U$.

12. Let H be a normal subgroup of the finite group G. Let P be a Sylow p-subgroup of H. Prove that $G = N_G(P).H$. (This is often called the Frattini argument.)

13. Let H be normal in G. Let P be a Sylow p-subgroup of G. Prove that $H \cap P$ is a Sylow p-subgroup of H.

CHAPTER 7.
GROUPS OF ORDERS 1 TO 15

7.1 Introduction

We have already found all possible groups of orders 1 to 5. In this chapter we realise our intention of finding all groups of orders up to 15. To do this we shall bring to bear the information and techniques discussed in the preceding chapters.

Throughout the following C_n will denote a cyclic group of order n. Isomorphic groups are identified.

7.2 Groups of Order 6

Let $|G| = 6 = 2 \times 3$. By Sylow's third theorem G has $1 + 3k$ Sylow 3-subgroups of order 3 and $(1 + 3k) \mid 6$, where k is one of $0, 1, 2, 3, \ldots$. Only $k = 0$ is possible. Hence there is just one Sylow 3-subgroup H which must be normal either by theorem 6.3.4 or because any subgroup of index 2 is normal by theorem 4.1.3.

G has $1 + 2k$ subgroups of order 2 and $(1 + 2k) \mid 6$. Thus G has 1 or 3 subgroups of order 2. If G has only one subgroup K of order 2, then by theorem 6.3.4 this subgroup is normal. In exactly the same way as in the proof of theorem 6.4.1 we have

$$H \cap K = \{e\},$$

$$H = \langle x \rangle ,$$

$$K = \langle y \rangle ,$$

$$xy = yx ,$$

and $\qquad G = \langle (xy) \rangle$

is cyclic.

By the structure theory for finitely generated abelian groups discussed in Chapter 5, this is the only abelian group of order 6.

The other possibility is that G is non-abelian with one normal subgroup of order 3 and 3 subgroups of order 2. Let H be the one normal subgroup of order 3. Then

$$H = \langle a \rangle = \{a, a^2, a^3 = e\},$$

with $O(a) = 3$.

Let $b \in G$, $b \notin H$. Now $|G/H| = |G|/|H| = 6/3 = 2$. Thus

$$G/H = \langle \bar{b} \rangle = \{\bar{b}, \bar{b}^2 = \bar{e}\},$$

where $\bar{e}$ is the coset H, and $\bar{b}$ is the coset $bH = Hb$. (Left and right cosets coincide because H is normal.) It follows that G must be the union of the cosets H and aH.

Now $H = \{a, a^2, a^3 = e\}$ and so $Hb = \{ab, a^2b, b\}$. $G = \{e, a, a^2, b, ab, a^2b\}$.

This means that if there is a non-abelian group of order 6 then its elements can be expressed in terms of two symbols a and b as above. We must now find how these elements multiply in order to obtain possible multiplication tables.

Now $ba \in bH = Hb = \{ab, a^2b, b\}$. Thus ba must be one of ab, a^2b, b. If $ba = b$, then $a = e$. If $ba = ab$, then G must be abelian. Neither of these two possibilities is correct. We conclude that $ba = a^2b$.

The order of b must divide $|G|$ and therefore must be 2 or 3. If $O(b) = 3$, then $\langle b \rangle$ is a subgroup of index 2 in G and therefore is normal in G. However, $\langle a \rangle$ is the only normal subgroup of G of order 3. We conclude that $O(b) = 2$. To sum up; there is at most one non-abelian group G of order 6. If such a group exists its elements can be written as:

$$\{e, a, a^2, b, ab, a^2b\},$$

where $ba = a^2b$, $O(a) = 3$, and $O(b) = 2$.

This determines a multiplication table identical with Table 2.2.1. We could show that this actually does define a group by checking the existence of an identity, an inverse for each element, and associativity. The latter is the only difficulty.

However, we already know that Table 2.2.1 determines the isomorphic groups D_3 and S_3.

Thus we have found a group isomorphic to these groups. In fact in this particular case we could have reached this conclusion immediately by noting that there is at most one non-abelian group of order 6 and that S_3 is such a group.

We could also have argued as follows. If a non-abelian group of order 6 does exist then our work shows that it must be described by:

$$\langle a,\ b \mid a^3 = b^2 = e,\ ba = a^2b,\ R \rangle ,$$

where R denotes that some more relations may be required. However, from the theory of presentation of groups in terms of generators and relations $\langle a,\ b \mid a^3 = b^2 = e,\ ba = a^2b \rangle$ is a group of order 6. Thus R is not needed and there is just one non-abelian group of order 6.

For a good introduction to the presentation of groups the reader is referred to: 'Presentation of Groups' — Johnson (C.U.P.). Other references are given at the end of this book.

The subgroups of S_3 are given in section 3.6. The conjugacy classes of S_3 are discussed in example 4.4.1. From these we deduce that $Z(S_3)$ is $\{(1)\}$. The commutator subgroup of S_3 is A_3. This can be calculated directly or deduced from theorem 4.5.4 and the facts that S_3/S_3' is abelian, S_3 is not abelian, and the only normal subgroups of S_3 are $\{(1)\}$, A_3, S_3.

For $S_3/A_3 \cong C_2$ see exercise 4.26. We have $S_3/\{(1)\} \cong S_3$ and $S_3/S_3 \cong \{e\}$. From the results obtained in section 3.6, S_3 has the lattice of subgroups as below:

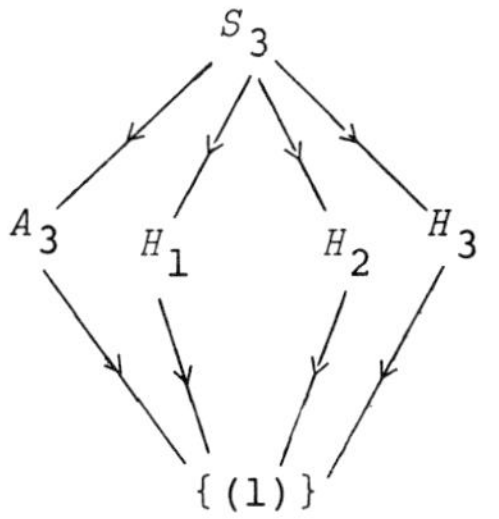

$$S_3 = \{(1),\ (12),\ (13),\ (23),\ (123),\ (132)\},$$
$$A_3 = \{(1),\ (123),\ (132)\} = S_3',$$
$$H_1 = \{(1),\ (12)\},\ H_2 = \{(1),\ (13)\},$$
$$H_3 = \{(1),\ (23)\},\ \{(1)\} = Z(S_3).$$

Figure 7.2.1

The cyclic group of order 6 has the lattice of subgroups given below:

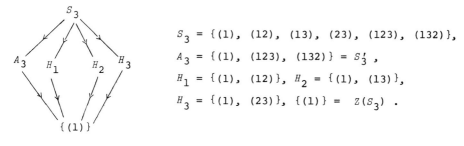

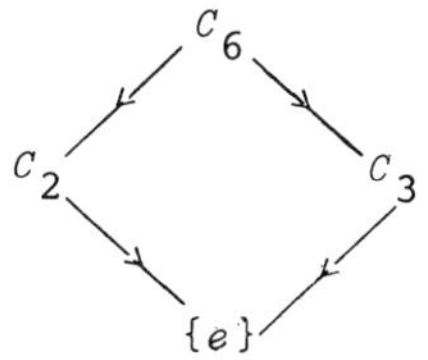

$$C_6 = \{e,\ a,\ a^2,\ a^3,\ a^4,\ a^5\} = \langle a \rangle$$
$$C_3 = \{e,\ a^2,\ a^4\} = \langle a^2 \rangle$$
$$C_2 = \{e,\ a^3\} = \langle a^3 \rangle$$

Figure 7.2.2

The conjugacy classes consist of single elements. All subgroups are normal.

$$Z(C_6) = C_6,$$
$$C_6' = \{e\},$$
$$C_6/C_2 \cong C_3,$$
$$C_6/C_3 \cong C_2,$$
$$C_6/C_6 \cong \{e\},$$
$$C_6/\{e\} \cong C_6.$$

7.3 Groups of Order 7

By theorem 3.1.3 there is upto isomorphism just one group of order 7. It is cyclic.

$$G = C_7 = \{e, a, a^2, a^3, a^4, a^5, a^6\} = \langle a \mid a^7 = e \rangle$$

C_7 has only trivial subgroups, namely C_7 and $\{e\}$.

The lattice of subgroups is:

C_7
|
$\{e\}$

$C_7/C_7 \cong \{e\}$, $C_7/\{e\} \cong C_7$,

$Z(C_7) = C_7$, $C_7' = \{e\}$.

Figure 7.3.1

The conjugacy classes consist of single elements.

7.4 Groups of Order 8

Let G be a group of order $8 = 2^3$. By the structure theory for finitely generated abelian groups there are 3 non-isomorphic abelian groups of order 8, namely:

$$C_2 \times C_2 \times C_2, \quad C_4 \times C_2, \quad C_8$$

Theorems 5.4.1 and 5.4.2 give the 3 possibilities and theorem 5.5.1 shows them to be non-isomorphic. That they are non-isomorphic can also be seen directly. $C_2 \times C_2 \times C_2$ contains no element of order 4 so cannot be isomorphic to $C_4 \times C_2$ which does. $C_2 \times C_2 \times C_2$ and $C_4 \times C_2$ contain no element of order 8 so cannot be isomorphic to C_8 which does.

In any abelian group G the conjugacy classes are single elements, all subgroups are normal, $Z(G) = G$ and $G' = \{e\}$. From now on we shall leave the reader to find the lattice of subgroups and the factor groups of the abelian groups of the remaining orders.

Now let G be non-abelian order 8. Since G has order 8, Lagrange's theorem 3.1.1 assures us that the elements of G must have orders 1, 2, 4, or 8. However, if $O(a) = 8$ then $G = \langle a \rangle$, which is abelian. Hence the elements of G must have orders 1, 2, or 4.

Suppose all elements $g \neq e$ of G have order 2. Let a, b be any 2 distinct elements of G. Then

$$a^2 = b^2 = (ab)^2 = e .$$

Thus $\quad abab = e \to a^2bab = a \to bab = a \to b^2ab = ba \to ab = ba.$

Hence G is abelian; in fact G must be $C_2 \times C_2 \times C_2$. Since we are supposing G is non-abelian it follows that not all elements of G have order 2. We conclude that at least one element $a \in G$ must have order 4. Then $\langle a \rangle = \{e, a, a^2, a^3\}$ is a cyclic subgroup of index 2 in G. By theorem 4.1.3 $H = \langle a \rangle$ is a normal subgroup of G.

Let $b \in G$, $b \notin \langle a \rangle$. Then

$$|G/H| = |G|/|H| = \frac{8}{4} = 2.$$

Thus

$$G/H = \langle \bar{b} \rangle = \{\bar{b}, \bar{b}^2 = \bar{e}\} ,$$

where $\bar{e}$ is the coset H, and $\bar{b}$ is the coset Hb, which is the same as bH since H is normal in G.

Thus G consists of the union of the cosets H and Hb. Since $H = \{e, a, a^2, a^3\}$ we have $Hb = \{b, ab, a^2b, a^3b\}$, and therefore $G = \{e, a, a^2, a^3, b, ab, a^2b, a^3b\}$.

At this stage we have shown that if G is a non-abelian group of order 8 then its elements can be described in terms of the symbols a and b by the above 8 words. Now we have to find the possible multiplication tables.

We consider ba. Since $ba \in bH = Hb$, it follows that ba must be one of b, ab, a^2b, a^3b. We consider the possibilities in turn. If $ba = b$, then $a = e$, which is not so.

If $ba = ab$, then G is abelian, which is not so.

If $ba = a^2b$, then $a = b^{-1}a^2b \to a^2 = (b^{-1}a^2b)(b^{-1}a^2b) \to a^2 = b^{-1}a^4b$

$$= b^{-1}eb = e , \quad \text{which is not so.}$$

We conclude that $ba = a^3b$.

Now consider b^2. Since $\overline{b^2} = \bar{b}^2 = \bar{e} = H$, we know that b^2 defines the coset H, that is $b^2 \in H$. Thus we have the possibilities:

$$b^2 = e,\ a,\ a^2,\ \text{or}\ a^3 .$$

If $b^2 = a$, then the order of b cannot be 2 otherwise $e = b^2 = a$ and cannot be 4 otherwise $e = b^4 = (b^2)^2 = a^2$. On the other hand, the order of every element $\neq e$ of the group G must be 2 or 4 as shown earlier. In particular $O(b) = 2$ or 4. From these contradictory results we conclude that $b^2 \neq a$.

If $b^2 = a^3$, then the order of b cannot be 2 otherwise $e = a^3$, and cannot be 4 otherwise $e = (b^2)^2 = (a^3)^2 = a^6$ which when combined with $a^4 = e$ gives $a^2 = e$. Again this contradicts the fact that $O(b) = 2$ or 4. We conclude that $b^2 \neq a^3$.

We are left with two possibilities:

$$b^2 = e,$$

$$b^2 = a^2 .$$

Case (1) $b^2 = e$.

$$G = \{e,\ a,\ a^2,\ a^3,\ b,\ ab,\ a^2b,\ a^3b\}$$

$$a^4 = e ,$$

$$b^2 = e ,$$

$$ba = a^3b .$$

This information suffices to determine the multiplication table 7.4.1 set out below.

As with the non-abelian group of order 6 it is easy to check that this table does have an identity e and an inverse for each element. Not so easy is the checking for associativity. We do this by exhibiting a known group with this multiplication table.

In section 1.1 we discussed the symmetry group of an equilateral triangle. This discussion can be repeated for any regular n-sided polygon, when it gives rise to a group called the dihedral group D_n of order $2n$. Let us consider D_4 in detail. D_4 is the symmetry group of the square.

	e	a	a^2	a^3	b	ab	a^2b	a^3b
e	e	a	a^2	a^3	b	ab	a^2b	a^3b
a	a	a^2	a^3	e	ab	a^2b	a^3b	b
a^2	a^2	a^3	e	a	a^2b	a^3b	b	ab
a^3	a^3	e	a	a^2	a^3b	b	ab	a^2b
b	b	a^3b	a^2b	ab	e	a^3	a^2	a
ab	ab	b	a^3b	a^2b	a	e	a^3	a^2
a^2b	a^2b	ab	b	a^3b	a^2	a	e	a^3
a^3b	a^3b	a^2b	ab	b	a^3	a^2	a	e

Table 7.4.1

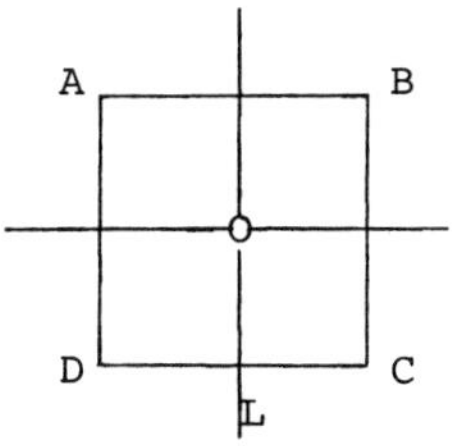

Figure 7.4.1

Let ABCD be a square with centre O. Let L be a line through the centre O fixed in space and parallel to a side of the square. Let a be a rotation through 90 degrees in an anti-clockwise sense about an axis through O perpendicular to the plane of the square. Let b be a rotation about the line L through 180 degrees.

All possible motions of the square which leave it apparently fixed in space can be described in terms of a and b. If e denotes the identity motion, we have $a^4 = e$, $b^2 = e$, and $ba = a^3b$, where ba means first carry out the motion a followed by the motion b. This gives 8 distinct symmetries of the square described by:

$$e,\ a,\ a^2,\ a^3,\ b,\ ab,\ a^2b,\ a^3b,$$

corresponding to the following final positions of the square after the motion written under the square has been carried out on the initial position.

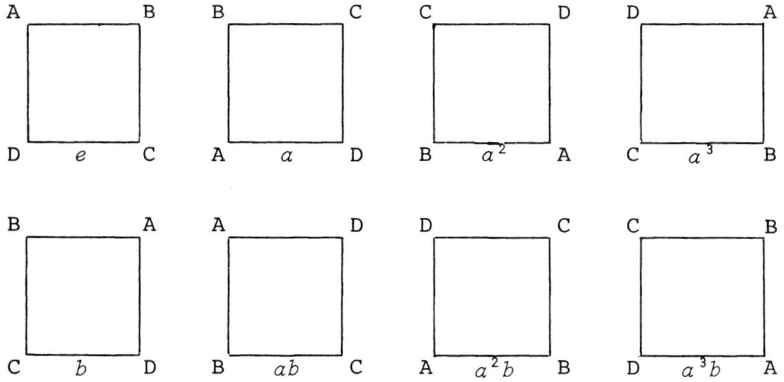

Figure 7.4.2

The multiplication table for this group can be obtained directly from the diagrams and the meaning of a and b. For example, $ab.a^2b$ means carry out the sequence of motions: first b, then a, then a, b, and finally a. Thus $ab.a^2b$ is equal to a^3.

Alternatively we can use the relations: $a^4 = b^2 = e$, $ba = a^3b$, to calculate $ab.a^2b = a(ba)ab = a(a^3b)ab = a^4(ba)b = a^4(a^3b)b = a^3$.

The multiplication table turns out to be exactly Table 7.4.1 that we obtained earlier as a possible group table for a group of order 8.

It follows that case (1) is the dihedral group D_4.

From the multiplication table, or directly from the symmetries of the square, we obtain the following information about D_4.

Element:	e	a	a^2	a^3	b	ab	a^2b	a^3b
Inverse:	e	a^3	a^2	a	b	ab	a^2b	a^3b
Order:	1	4	2	4	2	2	2	2

Conjugacy classes are:

$\{e\}$, $\{a^2\}$, $\{a, a^3\}$, $\{b, a^2b\}$, $\{ab, a^3b\}$.

To find the conjugacy classes it helps to remember that all elements in one class have the same order, and the number of elements in a class divides the order of the group. In fact, the number of elements in the class containing x is $[D_4 : N_{D_4}(x)]$. Moreover, only central elements lie in a class on their own. Thus the centre of D_4 is the union of the one element conjugacy classes. Hence $Z(D_4) = \{e, a^2\}$.

The commutator subgroup is:

$$D_4' = \langle \{x^{-1}y^{-1}xy \mid x, y \in D_4\} \rangle = \{e, a^2\} = Z(D_4) .$$

To find the other subgroups we note that the order of any subgroup divides the order of the group. In our case this means that possible subgroups are of orders 1, 2, 4, 8. $\{e\}$ is the subgroup of order 1.

By theorem 3.1.3 the subgroups of order 2 are cyclic and threfore are generated by the elements of order 2. This observation leads to the subgroups:

$D_4' = Z(D_4) = H_1 = \{e, a^2\}$, $H_2 = \{e, b\} = \langle b \rangle$,

$H_3 = \{e, ab\} = \langle ab \rangle$, $H_4 = \{e, a^2b\} = \langle a^2b \rangle$,

$H_5 = \{e, a^3b\} = \langle a^3b \rangle$.

By theorem 4.4.1, only H_1 is normal in D_4.

The subgroups of order 4 are either cyclic or Klein 4-groups of the form $\{e, x, y, xy = yx\} = K_4$ where $O(x) = O(y) = 2$. These considerations give the subgroups:

$H_6 = \{e, a, a^2, a^3\} = \langle a \rangle$, $H_7 = \{e, a^2, b, a^2b\} \cong K_4$

$\cong \{e, a^2, ab, a^3b\} = H_8$.

Since they have index 2 in D_4, the subgroups of order 4 are all normal in D_4. This also follows from the fact that they are unions of complete conjugacy classes (see theorem 4.4.1). The latter consideration also shows that we cannot have more than 3 subgroups of order 4. The only subgroup of order 8 is D_4 itself.

We have the lattice of subgroups:

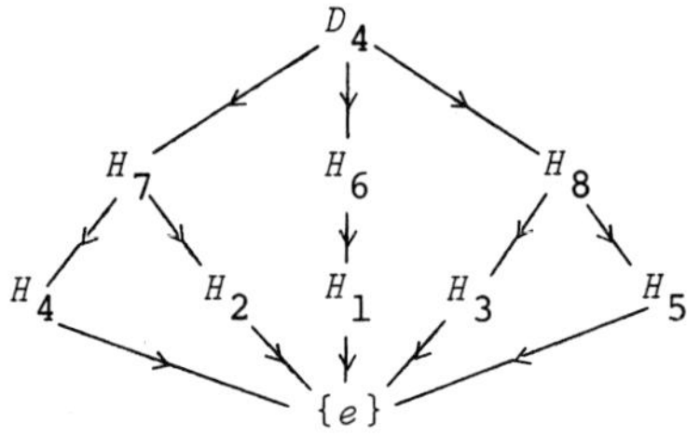

Figure 7.4.3

Factor groups are as follows:

$$D_4/H_6 \cong D_4/H_7 \cong D_4/H_8 \cong C_2.$$

To find $D_4/Z(D_4) = D_4/D_4'$, put $\bar{a} = aD_4'$, $\bar{b} = bD_4'$, $\bar{e} = D_4'$.

Note that $\bar{a}^2 = \overline{a^2} = \bar{e}$, because $a^2 \in Z(D_4) = D_4'$.

Also $(\bar{b})^2 = \overline{b^2} = \bar{e}$, because $b^2 = e \in D_4'$. However, $\bar{a} \neq \bar{e}$, since $a \notin Z(D_4) = D_4'$. Similarly $\bar{b} \neq \bar{e}$, because $b \notin Z(D_4)$. Moreover $[a, b] = a^{-1}b^{-1}ab \in D_4'$, hence

$$\overline{(a^{-1}b^{-1}ab)} = \bar{e} \ .$$

Therefore

$$(\bar{a})^{-1}(\bar{b})^{-1}\bar{a}\bar{b} = \bar{e} \ .$$

This means that

$$\bar{a}\bar{b} = \bar{b}\bar{a} \ .$$

Thus finally:

$$D_4/Z(D_4) = D_4/D_4' = \langle \bar{a}, \bar{b} \mid \bar{a}^2 = \bar{e}, \bar{b}^2 = \bar{e}, \bar{a}\bar{b} = \bar{b}\bar{a} \rangle$$
$$= \{\bar{e}, \bar{a}, \bar{b}, \bar{a}\bar{b} = \bar{b}\bar{a}\} \ ,$$

which is the Klein 4-group.

We turn now to

Case (2) $b^2 = a^2$.

In this case a possible group of order 8 is given by:

$$G = \{e,\ a,\ a^2,\ a^3,\ b,\ ab,\ a^2b,\ a^3b\}$$

with $a^4 = e$, $b^2 = a^2$, $ba = a^3b$; and we deduce $b^4 = e$. This information suffices to determine the following multiplication table.

	e	a	a^2	a^3	b	ab	a^2b	a^3b
e	e	a	a^2	a^3	b	ab	a^2b	a^3b
a	a	a^2	a^3	e	ab	a^2b	a^3b	b
a^2	a^2	a^3	e	a	a^2b	a^3b	b	ab
a^3	a^3	e	a	a^2	a^3b	b	ab	a^2b
b	b	a^3b	a^2b	ab	a^2	a	e	a^3
ab	ab	b	a^3b	a^2b	a^3	a^2	a	e
a^2b	a^2b	ab	b	a^3b	e	a^3	a^2	a
a^3b	a^3b	a^2b	ab	b	a	e	a^3	a^2

Table 7.4.2

As usual it is easy to verify that we have an identity e for the product defined by this table and an inverse for each element but not so easy to verify associativity. To do this we construct a system known to be a group which has the same multiplication table as above.

Let $a = \begin{pmatrix} 0 & i \\ i & 0 \end{pmatrix}$ and $b = \begin{pmatrix} 0 & 1 \\ -1 & 0 \end{pmatrix}$

where $i^2 = -1$. Then

$$a^2 = \begin{pmatrix} -1 & 0 \\ 0 & -1 \end{pmatrix}, \quad a^3 = \begin{pmatrix} 0 & -i \\ -i & 0 \end{pmatrix}, \quad ab = \begin{pmatrix} -i & 0 \\ 0 & i \end{pmatrix},$$

$$a^2b = \begin{pmatrix} 0 & -1 \\ 1 & 0 \end{pmatrix}, \quad a^3b = \begin{pmatrix} i & 0 \\ 0 & -i \end{pmatrix}, \quad e = \begin{pmatrix} 1 & 0 \\ 0 & 1 \end{pmatrix}.$$

This set of 8 matrices under ordinary matrix multiplication (see [G]) forms a group. This can be seen as follows. Matrix multiplication is associative and the multiplication table obtained by multiplying the given matrices is exactly Table 7.4.2. It follows that Table 7.4.2 does indeed define a group which is isomorphic to the group of 8 matrices just given. This group is called the quaternion group Q. (So called because of its close connection with a system of algebra called quaternion algebra invented by the Irish mathematician Hamilton around 1850 to attack certain problems in geometry.)

From the multiplication table, or from the description

$$\langle a, b \mid a^4 = e, b^2 = a^2, ba = a^3b \rangle ,$$

we obtain the following information about Q.

Element:	e	a	a^2	a^3	b	ab	a^2b	a^3b
Inverse:	e	a^3	a^2	a	a^2b	a^3b	b	ab
Order:	1	4	2	4	4	4	4	4

The conjugacy classes are:

$\{e\}$, $\{a^2\}$, $\{a, a^3\}$, $\{b, a^2b\}$, $\{ab, a^3b\}$.

The centre $Z(Q) = \{e, a^2\}$ = commutator subgroup Q'. Subgroups are of possible orders 1, 2, 4, 8. $\{e\}$ is the subgroup of order 1. Subgroups of order 2 are cyclic, hence there is only one generated by a^2, namely $H_1 = Z(Q) = Q'$. H_1 is normal.

Subgroups of order 4 are cyclic or Klein 4-groups. The latter are generated by two elements each of order 2. There is only one element of order 2 in Q. Thus all subgroups of order 4 are cyclic. Hence the subgroups of order 4 are:

$H_2 = \{e, a, a^2, a^3\} = \langle a \rangle = \langle a^3 \rangle$,
$H_3 = \{e, b, a^2, a^2b\} = \langle b \rangle = \langle a^2b \rangle$,
$H_4 = \{e, ab, a^2, a^3b\} = \langle ab \rangle = \langle a^3b \rangle$.

As subgroups of index 2 in Q, or as unions of complete conjugacy classes, these subgroups of order 4 are all normal. The only subgroup of order 8 is Q itself.

The lattice of subgroups is:

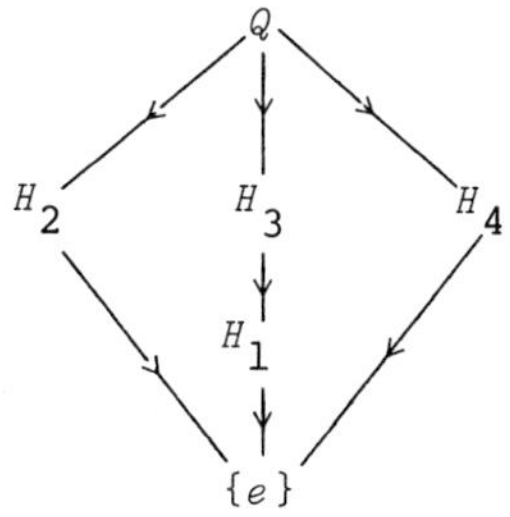

Figure 7.4.4

Factor groups are:

$$Q/H_2 \cong Q/H_3 \cong Q/H_4 \cong C_2 .$$

$Q/H_1 = Q/Q' = Q/Z(Q)$ is generated by $\bar{a} = aQ'$ and $\bar{b} = bQ'$. A similar discussion to that for D_4/D_4' shows that

$$Q/H_1 \cong \langle \bar{a}, \bar{b} \mid \bar{a}^2 = \bar{b}^2 = \bar{e}, \bar{a}\bar{b} = \bar{b}\bar{a} \rangle ,$$

which is the Klein 4-group.

Note that all subgroups of Q other than Q itself (i.e. all proper subgroups) are cyclic and normal.

7.5 Groups of Order 9

Let G be a group of order $9 = 3^2$. By theorem 4.9.1 all groups of order 9 are abelian. By the structure theory for finitely generated abelian groups, theorems 5.4.1 and 5.4.2,

$$G = C_9 \text{ or } C_3 \times C_3 .$$

By theorem 5.5.1 these are non-isomorphic. Directly, $C_3 \times C_3$ has no element of order 9 but C_9 has. Hence there are two groups of order 9, both abelian.

7.6 Groups of Order 10

Let G be a group of order $10 = 2 \times 5$. By theorems 5.4.1 and 5.4.2, the only abelian group of order 10 is $C_2 \times C_5 = C_{10}$.

By Sylow's third theorem each non-abelian group of order 10 has just one subgroup of order 5 which is necessarily normal, and $1 + 2k$ subgroups of order 2. Since $(1 + 2k)$ must divide 10, and $(1 + 2k) \neq 1$, otherwise the group is abelian by exactly the same argument as in the proof of theorem 6.4.1, we conclude that there are 5 subgroups of order 2.

Let H be the **normal** subgroup of order 5. Let K be **one** of the subgroups of order 2. We have

$$H = \{e, a, a^2, a^3, a^4\} = \langle a \rangle$$

and

$$K = \{e, b\} = \langle b \rangle .$$

By the usual argument

$$H \cap K = \{e\} .$$

Hence $b \notin H$. Now

$$|G/H| = |G|/|H| = 2 .$$

Hence

$$G/H = \{\bar{e}, \bar{b}\} .$$

It follows that

$$G = H \cup Hb = \{e, a, a^2, a^3, a^4, b, ab, a^2b, a^3b, a^4b\} .$$

Now $ba \in bH = Hb$. Thus ba is one of:

$$b,\ ab,\ a^2b,\ a^3b,\ a^4b\ .$$

We check each possibility in turn and conclude that

$$ba = a^4b\ .$$

Thus

$$G = \{e,\ a,\ a^2,\ a^3,\ a^4,\ b,\ ab,\ a^2b,\ a^3b,\ a^4b\}\ ,$$

with $ba = a^4b$. This suffices to determine the multiplication table 7.6.1.

	e	a	a^2	a^3	a^4	b	ab	a^2b	a^3b	a^4b
e	e	a	a^2	a^3	a^4	b	ab	a^2b	a^3b	a^4b
a	a	a^2	a^3	a^4	e	ab	a^2b	a^3b	a^4b	b
a^2	a^2	a^3	a^4	e	a	a^2b	a^3b	a^4b	b	ab
a^3	a^3	a^4	e	a	a^2	a^3b	a^4b	b	ab	a^2b
a^4	a^4	e	a	a^2	a^3	a^4b	b	ab	a^2b	a^3b
b	b	a^4b	a^3b	a^2b	ab	e	a^4	a^3	a^2	a
ab	ab	b	a^4b	a^3b	a^2b	a	e	a^4	a^3	a^2
a^2b	a^2b	ab	b	a^4b	a^3b	a^2	a	e	a^4	a^3
a^3b	a^3b	a^2b	ab	b	a^4b	a^3	a^2	a	e	a^4
a^4b	a^4b	a^3b	a^2b	ab	b	a^4	a^3	a^2	a	e

Table 7.6.1

As usual the existence of an identity and an inverse for each element follows from the table.

To prove associativity we consider D_5 the symmetry group of a regular plane pentagon ABCDE.

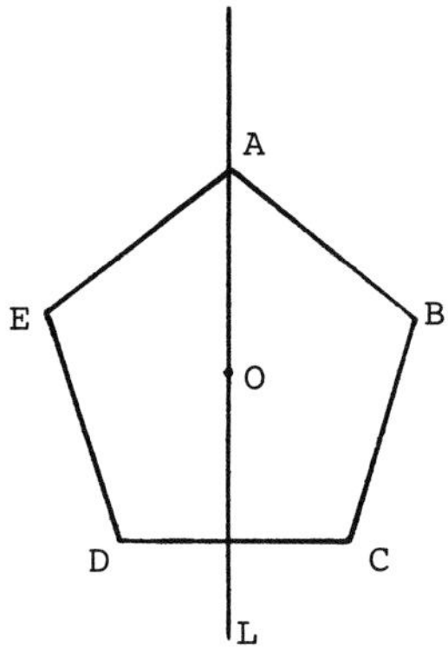

Figure 7.6.1

Let ABCDE be a regular pentagon with centre O. Let L be a line through the centre O fixed in space and perpendicular to a side of the pentagon. Let a be a rotation through 72 degrees in an anti-clockwise sense about an axis through O perpendicular to the plane of the pentagon. Let b be a rotation about the line L through 180 degrees.

All possible motions of the pentagon which leave it apparently fixed in space can be described in terms of a and b. The working is now exactly as for the symmetry group of the square considered earlier. We obtain a group whose multiplication table is Table 7.6.1.

Note ab means: first b then a.

It follows that the dihedral group D_5 is the only non-abelian group of order 10.

From the multiplication table, or directly from the symmetries of the pentagon, or even from the description:

$$\langle a,\ b \mid a^5 = e,\ b^2 = e,\ ba = a^4b \rangle ,$$

we obtain the following information about D_5.

Element:	e	a	a^2	a^3	a^4	b	ab	a^2b	a^3b	a^4b
Inverse:	e	a^4	a^3	a^2	a	b	ab	a^2b	a^3b	a^4b
Order:	1	5	5	5	5	2	2	2	2	2

The conjugacy classes are:

$$\{e\},\ \{a, a^4\},\ \{a^2, a^3\},\ \{b, ab, a^2b, a^3b, a^4b\}\ .$$

The centre $Z(D_5) = \{e\}$. Subgroups are of possible orders 1, 2, 5, 10. $\{e\}$ is the subgroup of order 1. Subgroups of order 2 are cyclic. By Sylow's third theorem there are 5 of them. They are as follows:

$H_1 = \{e, b\} = \langle b \rangle$, $H_2 = \{e, ab\} = \langle ab \rangle$, $H_3 = \{e, a^2b\} = \langle a^2b \rangle$,

$H_4 = \{e, a^3b\} = \langle a^3b \rangle$, $H_5 = \{e, a^4b\} = \langle a^4b \rangle$.

Subgroups of order 5 are also cyclic. By Sylow's third theorem there is only one which has to be normal. It is $H_6 = \langle a \rangle$. Since D_5/H_6 is of order 2 hence cyclic, $H_6 \supset D_5'$ by theorem 4.5.4. However, $D_5' \neq \{e\}$ since D_5 is non-abelian. Finally, since D_5' is normal and the only normal subgroups of D_5 are $\{e\}$, H_6, D_5, it follows that $H_6 = D_5'$. Thus

$$D_5' = \{e, a, a^2, a^3, a^4\} = \langle a \rangle\ .$$

The only subgroup of order 10 is D_5 itself.

The lattice of subgroups is:

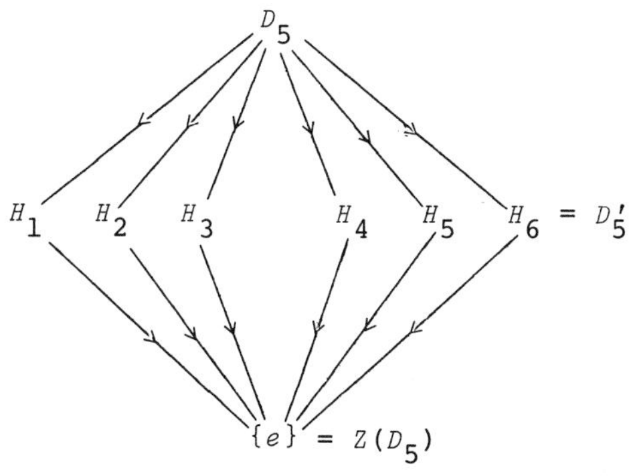

Figure 7.6.2

Factor groups are:

the trivial $D_5/D_5 \cong \{e\}$, $D_5/\{e\} \cong D_5$ and $D_5/D_5' \cong C_2$.

7.7 **Groups of Order 11**

By theorem 3.1.3 the only group of order 11 is C_{11}. The lattice of subgroups is:

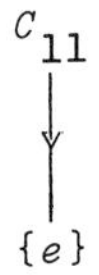

Figure 7.7.1

7.8 **Groups of Order 12**

Let G be a group of order $12 = 2^2 \times 3$. By the structure theory for abelian groups we deduce that the abelian groups of order 12 are:

$$C_{12}, \; C_4 \times C_3, \quad C_2 \times C_2 \times C_2 \quad .$$

The usual considerations about the orders of elements or an appeal to the uniqueness theorem 5.5.1, shows that these groups are non-isomorphic.

Now let G be non-abelian of order 12. By Sylow's third theorem G must have $s_2 = 1 + 2k$ subgroups of order $2^2 = 4$ and $s_3 = 1 + 3\ell$ subgroups of order 3, where both $1 + 2k$ and $1 + 3\ell$ must divide 12. Hence $s_2 = 1$ or 3 and $s_3 = 1$ or 4.

If $s_3 = 4$, let H be any of the 4 subgroups of order 3 and let K be either a subgroup of order 4 or a subgroup of order 3 **not** H.

$|H \cap K|$ divides $|H|$ and $|K|$ by Lagrange's theorem 3.1.1. Hence

$$|H \cap K| = 1.$$

Thus $$H \cap K = \{e\} \; .$$

Therefore the 4 subgroups of order 3 together with 1 subgroup of order 4 account for $4 \times (3 - 1) + 3 + 1 = 12$ elements of G. But there are only 12 elements in G. Hence if $s_3 = 4$ then s_2 must be 1.

This means that we have 3 possibilities.

(1) $s_2 = 1$ $s_3 = 1$

(2) $s_2 = 1$ $s_3 = 4$

(3) $s_2 = 3$ $s_3 = 1$

Case (1) $s_2 = 1$ $s_3 = 1$

By theorem 6.3.4 G has a normal subgroup K of order $2^2 = 4$ and a normal subgroup H of order 3. The usual argument runs: as $|H \cap K|$ divides $|H|$ and $|K|$ we have $H \cap K = \{e\}$. Let $h \in H$ and $k \in K$. Then $[h, k] = h^{-1}(k^{-1}hk) \in H$ because H is normal. Similarly $[h, k] = (h^{-1}k^{-1}h)k \in K$ because K is normal. Thus

$$[h, k] \in H \cap K = \{e\} .$$

Hence $h^{-1}k^{-1}hk = e$.

Thus $hk = kh$.

By theorem 3.4.2 (3)

$$|HK| = |H||K|/|H \cap K| = 12 = |G| .$$

Thus $G = HK$.

Moreover H of order 3 and K of order 4 must both be abelian. Altogether we conclude that $G = HK$ is abelian. This case has been considered already. However, it is of interest to note that we can argue as follows.

H must be cyclic C_3 and K must be cyclic C_4 or the Klein 4-group. The latter may be written $C_2 \times C_2$. Since the above work shows that $G = H \times K$, we have $G = C_3 \times C_4$ or $C_3 \times C_2 \times C_2$. This is the result found already, but we have avoided using the structure theory of Chapter 5.

Case (2) $s_2 = 1$ $s_3 = 4$

Let K be the normal subgroup of order 4 which exists by theorem 6.3.4. G/K has order $12/4 = 3$. Thus $G/K \cong C_3$. Hence for any $b \notin K$ we have

$$G/K = \{\bar{e}, \bar{b}, \bar{b}^2\} \cong C_3 .$$

Let H be any Sylow 3-subgroup order 3. As above $H \cap K = \{e\}$. This means that we can choose b to be any element of H not equal to e.

Let $b^3 = e$ so that $H = \langle b \rangle$. We have

$$\bar{e} = K,\ \bar{b} = bK,\ \bar{b}^2 = \overline{b^2} = b^2 K ,$$

and $$G = K \cup bK \cup b^2 K .$$

Now K is of order 4 and therefore may be C_4 or K_4, the Klein 4-group.

(a) First take $K = C_4 = \{e, a, a^2, a^3\} = \langle a \mid a^4 = e \rangle$.

Then $$G = \{e, a, a^2, a^3, b, ba, ba^2, ba^3, b^2, b^2a, b^2a^2, b^2a^3\}.$$

Consider ab. Now $ab \in Kb$. However, $Kb = bK$, because K is normal in G. Thus

$$ab \in bK .$$

Hence ab is one of b, ba, ba^2, ba^3.

As usual we check each possibility in turn and conclude that ab must be equal to ba^3. But then

$$(ba)^2 = b(ab)a = b(ba^3)a = b^2a^4 = b^2 .$$

Hence $$(ba)^3 = (ba)^2 ba = b^2.ba = b^3 a = a .$$

Also from $b^3 = e$ we have $b^{-1} = b^2 = (ba)^2$. Thus

$$b^{-1} = b^2 \in \langle (ba) \rangle .$$

Hence $$b \in \langle (ba) \rangle ,$$

and $$a = (ba)^3 \in \langle (ba) \rangle .$$

Thus $G \subset \langle (ba) \rangle$. However, $\langle (ba) \rangle \subset G$. It follows that if G is a group then $G = \langle (ba) \rangle$ is cyclic and therefore abelian.

We conclude that there is <u>no</u> non-abelian group in which $s_2 = 1$, $s_3 = 4$, and the Sylow subgroup order 4 is C_4.

(b) Now take $K = K_4 = \{e, a, c, ac = ca\}$

$$= \langle a, c \mid a^2 = c^2 = (ac)^2 = e \rangle ,$$

the Klein 4-group.

As above $G = K \cup bK \cup b^2K$. Thus

$$G = \{e,\ a,\ c,\ ac,\ b,\ ba,\ bc,\ bac,\ b^2,\ b^2a,\ b^2c,\ b^2ac\} .$$

Now K is normal in G. Hence

$$b^{-1}Kb = K.$$

As G is non-abelian, for at least one $x \in K$, we have

$$b^{-1}xb \neq x .$$

Put $b^{-1}xb = y \in K$. Since K is generated by any two of its elements $\neq e$ with the same defining relations, we obtain the same G, upto isomorphism, no matter which two elements x and y we choose from K. Thus take $x = a$ and $y = c$. Then $b^{-1}ab = c$. Thus

$$ab = bc .$$

We now require $b^{-1}cb$. We know from $b^{-1}Kb = K$ that $b^{-1}cb \in K$. Hence $b^{-1}cb$ is one of: $e,\ a,\ c,\ ac$.

We check each possibility and conclude that

$$b^{-1}cb = ac .$$

That is, $cb = bac$.

From this we deduce that

$$acb = ba .$$

Hence if there is a group G of order 12 with $s_2 = 1$ and $s_3 = 4$ it must consist of the elements:

$\{e,\ a,\ c,\ ac,\ b,\ ba,\ bc,\ bac,\ b^2,\ b^2a,\ b^2c,\ b^2ac\}$,

which multiply according to the rules:

$a^2 = c^2 = e,\ b^3 = e,\ ac = ca,\ ab = bc,\ cb = bac,\ acb = ba$.

Each of the last four rules enables us to 'turn round' the basic elements $a,\ c,\ ac,\ b$ if they arise in the 'wrong order'. For example $b.ac$ is one of the listed elements so $b.ac$ is in the 'right order'. On the other hand, $ac.b$ is in the 'wrong order'. However, we can turn the elements round by using $ac.b = ba$, which is a listed element. It follows that these rules suffice to calculate the multiplication table.

	e	a	c	ac	b	ba	bc	bac	b^2	b^2a	b^2c	b^2ac
e	e	a	c	ac	b	ba	bc	bac	b^2	b^2a	b^2c	b^2ac
a	a	e	ac	c	bc	bac	b	ba	b^2ac	b^2c	b^2a	b^2
c	c	ac	e	a	bac	bc	ba	b	b^2a	b^2	b^2ac	b^2c
ac	ac	c	a	e	ba	b	bac	bc	b^2c	b^2ac	b^2	b^2a
b	b	ba	bc	bac	b^2	b^2a	b^2c	b^2ac	e	a	c	ac
ba	ba	b	bac	bc	b^2c	b^2ac	b^2	b^2a	ac	c	a	e
bc	bc	bac	b	ba	b^2ac	b^2c	b^2a	b^2	a	e	ac	c
bac	bac	bc	ba	b	b^2a	b^2	b^2ac	b^2c	c	ac	e	a
b^2	b^2	b^2a	b^2c	b^2ac	e	a	c	ac	b	ba	bc	bac
b^2a	b^2a	b^2	b^2ac	b^2c	c	ac	e	a	bac	bc	ba	b
b^2c	b^2c	b^2ac	b^2	b^2a	ac	c	a	e	ba	b	bac	bc
b^2ac	b^2ac	b^2c	b^2a	b^2	a	e	ac	c	bc	bac	b	ba

Table 7.8.1

From the table it follows that there is an identity and an inverse for each element. As usual associativity is checked by producing a known group with this multiplication table.

Consider A_4 the alternating group on 4 symbols (see exercise 3.11). The elements of A_4 can be written:

$e = (1)$, $a = (12)(34)$, $c = (13)(24)$, $ac = (14)(23)$,

$b = (123)$, $ba = (134)$, $bc = (243)$, $bac = (142)$,

$b^2 = (132)$, $b^2a = (234)$, $b^2c = (124)$, $b^2ac = (143)$.

In terms of a, b, c the multiplication table of A_4 coincides with Table 7.8.1.

We conclude that A_4 is the only group of order 12 with $s_2 = 1$ and $s_3 = 4$. (**Note** a simple description of A_4 exists in exercise 7.5.)

From the multiplication table and its representation in terms of A_4, we deduce the following information about A_4.

Element:	e	a	c	ac	b	ba	bc	bac	b^2	b^2a	b^2c	b^2ac
Inverse:	e	a	c	ac	b^2	b^2ac	b^2a	b^2c	b	bc	bac	ba
Order:	1	2	2	2	3	3	3	3	3	3	3	3

To help find the conjugacy classes we use:

(1) exercise 4.6, conjugate permutations must have the same cycle pattern, hence the same order;

(2) the number of elements in a conjugacy class divides the order of the group.

The conjugacy classes are:

$\{(1)\}$, $\{(12)(34), (13)(24), (14)(23)\}$,

$\{(123), (142), (134), (243)\}$, $\{(132), (124), (143), (234)\}$.

The centre of A_4 is the union of the one-element conjugacy classes. Hence

$$Z(A_4) = \{(1)\} .$$

To find the commutator subgroup we argue as follows. The subgroups are of possible orders 1, 2, 3, 4, 6, 12. However, 6 is ruled out according to exercise 4.16. Let $\{e, x\}$ be a typical subgroup of order 2. Suppose it to be normal in A_4. Then $g^{-1}xg = x$ for all $g \in A_4$. Thus

$$x \in Z(A_4) .$$

Thus $\quad x = e$. Contradiction.

It follows that **no** subgroup of order 2 is normal in A_4. Thus

$$|A_4'| \neq 2 \ .$$

Now A_4 is not abelian, hence

$$A_4' \neq \{e\}$$

Thus $\quad |A_4'| = 3, 4 \text{ or } 12.$

However, there are $s_3 = 4$ subgroups of order 3. Hence by theorem 6.3.4 **no** subgroup of order 3 is normal. Thus

$$|A_4'| = 4 \text{ or } 12 \ .$$

Now

$$K_4 = \{(1),\ (12)(34),\ (13)(24),\ (14)(23)\} = \{e,\ a,\ c,\ ac\} = K \ ,$$

is normal in A_4. Moreover $|A_4/K| = 3$, hence

$$A_4/K \cong C_3 \ ,$$

which is abelian. By theorem 4.5.4, $A_4' \subset K$ with $|K| = 4$. Thus

$$A_4' = K \ .$$

Hence $\quad A_4' = \{(1),\ (12)(34),\ (13)(24),\ (14)(23)\}.$

Subgroup order 1 is: $\{(1)\} = \{e\}$.

Subgroups order 2 are: $H_1 = \{e,\ a\} = \langle a \rangle$, $H_2 = \{e,\ c\} = \langle c \rangle$, $H_3 = \{e,\ ac\} = \langle ac \rangle$.

Subgroups order 3 are: $H_4 = \{e,\ b,\ b^2\} = \langle b \rangle$,

$H_5 = \{e,\ bc,\ b^2 a\} = \langle bc \rangle$,

$H_6 = \{e,\ ba,\ b^2 ac\} = \langle ba \rangle$,

$H_7 = \{e,\ bac,\ b^2 c\} = \langle bac \rangle$.

Since $s_2 = 1$ the only subgroup of order 4 is: $A_4' = H_8$. By exercise 4.16 there are no subgroups order 6. The only subgroup order 12 is A_4 itself.

The lattice of subgroups is:

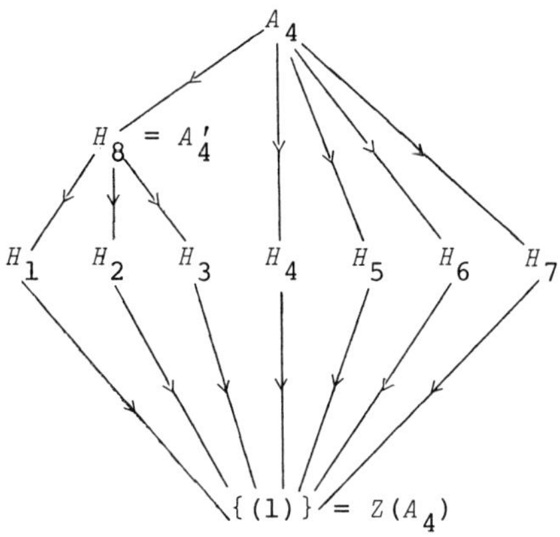

Figure 7.8.1

Factor groups are: $A_4/\{(1)\} \cong A_4$, $A_4/A_4' \cong C_3$, $A_4/A_4 \cong \{e\}$.

<u>**Case (3)**</u> $s_2 = 3$ $s_3 = 1$.

Let $H = \{e, b, b^2\} = \langle b \mid b^3 = e \rangle$ be the normal subgroup of G of order 3.

Let K be <u>**any one**</u> of the three <u>**isomorphic**</u> subgroups of G of order 4. By the usual argument $H \cap K = \{e\}$. K must be either C_4 or K_4, the Klein 4-group.

(a) First take $K = C_4 = \{e, a, a^2, a^3\} = \langle a \mid a^4 = e \rangle$.

Since $H \cap K = \{e\}$, we have $a^i \notin H$ for $i = 1, 2, 3$. Let $\bar{e} = H$, $\bar{a} = aH$. Then $\bar{a}^i = \overline{a^i} \neq \bar{e}$ for $i = 1, 2, 3$. Since $|G/H| = |G|/|H| = 4$, it follows that

$$G/H = \langle \bar{a} \mid \bar{a}^4 = \bar{e} \rangle = \{\bar{e}, \bar{a}, \bar{a}^2, \bar{a}^3\} .$$

Hence $G = H \cup aH \cup a^2H \cup a^3H$.

Thus $G = \{e, b, b^2, a, ab, ab^2, a^2, a^2b, a^2b^2, a^3, a^3b, a^3b^2\}$.

Consider ba. We have $ba \in Ha$.

Now H is normal in G. Hence $Ha = aH$. Thus

$$ba \in aH = \{a, ab, ab^2\} .$$

Hence ba is one of a, ab, ab^2. If $ba = a$, then $b = e$, which is not true. If $ba = ab$, then G is abelian, which is not true. Hence

$$ba = ab^2.$$

Altogether we have:

$G = \{e, b, b^2, a, ab, ab^2, a^2, a^2b, a^2b^2, a^3, a^3b, a^3b^2\}$,

with $a^4 = e$, $b^3 = e$, $ba = ab^2$.

These suffice to determine the multiplication table 7.8.2.

	e	b	b^2	a	ab	ab^2	a^2	a^2b	a^2b^2	a^3	a^3b	a^3b^2
e	e	b	b^2	a	ab	ab^2	a^2	a^2b	a^2b^2	a^3	a^3b	a^3b^2
b	b	b^2	e	ab^2	a	ab	a^2b	a^2b^2	a^2	a^3b^2	a^3	a^3b
b^2	b^2	e	b	ab	ab^2	a	a^2b^2	a^2	a^2b	a^3b	a^3b^2	a^3
a	a	ab	ab^2	a^2	a^2b	a^2b^2	a^3	a^3b	a^3b^2	e	b	b^2
ab	ab	ab^2	a	a^2b^2	a^2	a^2b	a^3b	a^3b^2	a^3	b^2	e	b
ab^2	ab^2	a	ab	a^2b	a^2b^2	a^2	a^3b^2	a^3	a^3b	b	b^2	e
a^2	a^2	a^2b	a^2b^2	a^3	a^3b	a^3b^2	e	b	b^2	a	ab	ab^2
a^2b	a^2b	a^2b^2	a^2	a^3b^2	a^3	a^3b	b	b^2	e	ab^2	a	ab
a^2b^2	a^2b^2	a^2	a^2b	a^3b	a^3b^2	a^3	b^2	e	b	ab	ab^2	a
a^3	a^3	a^3b	a^3b^2	e	b	b^2	a	ab	ab^2	a^2	a^2b	a^2b^2
a^3b	a^3b	a^3b^2	a^3	b^2	e	b	ab	ab^2	a	a^2b^2	a^2	a^2b
a^3b^2	a^3b^2	a^3	a^3b	b	b^2	e	ab^2	a	ab	a^2b	a^2b^2	a^2

Table 7.8.2

As usual we verify that this actually is the multiplication table of a group by producing a known group with Table 7.8.2 as its multiplication table.

$$\text{Let } a = \begin{pmatrix} 0 & i \\ i & 0 \end{pmatrix}, \quad b = \begin{pmatrix} \omega & 0 \\ 0 & \omega^2 \end{pmatrix},$$

where $i^2 = -1$, $\omega = -\frac{1}{2} + i\frac{\sqrt{3}}{2}$, $\omega^2 = -\frac{1}{2} - i\frac{\sqrt{3}}{2}$, $\omega^3 = 1$.

By direct calculation we have:

$$a^2 = \begin{pmatrix} -1 & 0 \\ 0 & -1 \end{pmatrix}, \quad b^2 = \begin{pmatrix} \omega^2 & 0 \\ 0 & \omega \end{pmatrix}, \quad a^3 = \begin{pmatrix} 0 & -i \\ -i & 0 \end{pmatrix}, \quad a^4 = b^3 = \begin{pmatrix} 1 & 0 \\ 0 & 1 \end{pmatrix},$$

$$ab = \begin{pmatrix} 0 & i\omega^2 \\ i\omega & 0 \end{pmatrix}, \quad a^2b = \begin{pmatrix} -\omega & 0 \\ 0 & -\omega^2 \end{pmatrix}, \quad a^3b = \begin{pmatrix} 0 & -i\omega^2 \\ -i\omega & 0 \end{pmatrix},$$

$$ab^2 = \begin{pmatrix} 0 & i\omega \\ i\omega^2 & 0 \end{pmatrix}, \quad a^2b^2 = \begin{pmatrix} -\omega^2 & 0 \\ 0 & -\omega \end{pmatrix}, \quad a^3b^2 = \begin{pmatrix} 0 & -i\omega \\ -i\omega^2 & 0 \end{pmatrix}.$$

The reader should check that the 12 matrices above under ordinary matrix multiplication (see [G]) form a group M with the multiplication table 7.8.2.

We conclude that there is just one group of order 12 upto isomorphism with one subgroup of order 3, necessarily normal, and 3 cyclic subgroups of order 4, namely the group M.

From Table 7.8.2, the relations $a^4 = e$, $b^3 = e$, $ba = ab^2$, and the considerations below we find the following information about M.

Element:	e	b	b^2	a	ab	ab^2	a^2	a^2b	a^2b^2	a^3	a^3b	a^3b^2
Inverse:	e	b^2	b	a^3	a^3b	a^3b^2	a^2	a^2b^2	a^2b	a	ab	ab^2
Order:	1	3	3	4	4	4	2	6	6	4	4	4

The conjugacy classes are:

$\{e\}$, $\{a^2\}$, $\{b, b^2\}$, $\{a^2b, a^2b^2\}$, $\{a, ab, ab^2\}$, $\{a^3, a^3b, a^3b^2\}$.

The centre of M is the union of the single element conjugacy classes. Hence $Z(M) = \{e, a^2\} = \langle a^2 \rangle$. Since $|M/H| = 4$, M/H is abelian. By theorem 4.5.4, $H \supset M'$. Thus $|M'|$ divides $|H| = 3$. Hence $|M'| = 1$ or 3. However, M is **not** abelian. Hence

$$M' \neq \{e\} .$$

Thus $|M'| = 3$. Hence

$$M' = H = \{e, b, b^2\} = \langle b \rangle \quad .$$

Thus the commutator subgroup of M is:

$$M' = \{e, b, b^2\} = \langle b \mid b^3 = e \rangle \; .$$

Subgroups are of possible orders 1, 2, 3, 4, 6, 12. Those of order 2 are cyclic. As mentioned above, there are $s_3 = 1$ subgroup of order 3 and $s_2 = 3$ **cyclic** subgroups of order 4. Any subgroup of order 6 is of index 2 in M and hence is normal. As such, it is the union of conjugacy classes. Moreover, no element of order 4 can lie in it because $4 \nmid 6$. These considerations give the following subgroups.

Subgroup order 1: $\{e\}$.

Subgroups order 2: $H_1 = \{e, a^2\} = \langle a^2 \rangle = Z(M)$.

Subgroups order 3: $H_2 = \{e, b, b^2\} = \langle b \rangle = H = M'$.

Subgroups order 4: $H_3 = \{e, a, a^2, a^3\} = \langle a \rangle = K$,

$H_4 = \{e, ab, a^2, a^3b\} = \langle ab \rangle$, $H_5 = \{e, ab^2, a^2, a^3b^2\} = \langle ab^2 \rangle$.

Subgroups order 6: $H_6 = \{e, a^2b, b^2, a^2, b, a^2b^2\} = \langle a^2b \rangle$, which is **normal**.

The lattice of subgroups is:

(**Note** that all **proper** subgroups are cyclic.)

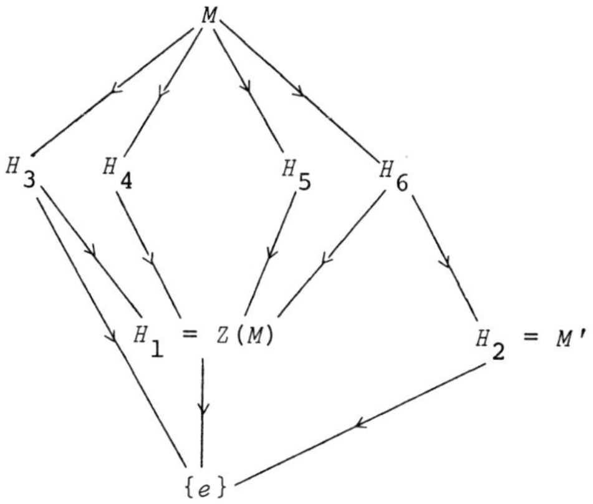

Figure 7.8.2

Factor groups are:

$M/Z(M) = \langle \bar{a}, \bar{b} \rangle$ is of order 6 where $\bar{a} = aZ(M)$, $\bar{b} = bZ(M)$.

Now $\bar{a}\bar{b} = \bar{b}\bar{a}$ if and only if $\overline{(a^{-1}b^{-1}ab)} = \bar{e}$. This is so if and only if $a^{-1}b^{-1}ab \in Z(M)$. However, $a^{-1}b^{-1}ab = a^3b^2ab = b^2 \notin Z(M)$. Thus

$$\bar{a}\bar{b} \neq \bar{b}\bar{a}.$$

Hence $M/Z(M)$ is not abelian. Hence

$$M/Z(M) \cong S_3.$$

In fact

$$M/Z(M) = \langle \bar{a}, \bar{b} \mid \bar{a}^2 = \bar{e}, \bar{b}^3 = \bar{e}, \bar{b}\bar{a} = \bar{a}\bar{b}^2 \rangle$$

$$\cong S_3 .$$

$$M/M' = \langle \bar{a}, \bar{b} \rangle,$$

where $\bar{a} = aM'$, $\bar{b} = bM'$. However,

$$\bar{b} = bM' = M' = \bar{e} .$$

Thus $\quad M/M' = \langle \bar{a} \rangle = \{\bar{e}, \bar{a}, \bar{a}^2, \bar{a}^3\} \cong C_4$.

Finally

$M/H_6 \cong C_2$, $M/\{e\} \cong M$, and $M/M \cong \{e\}$.

(b) Let $K = \{e, a, c, ac\} = \langle a, c \mid a^2 = c^2 = (ac)^2 = e \rangle = K_4$ the Klein 4-group. (Since this is a Sylow 4-subgroup and all such groups are conjugate and therefore isomorphic, it follows that all subgroups of G of order 4 are Klein 4-groups.)

As before,

$$H = \{e, b, b^2\} = \langle b \mid b^3 = e \rangle.$$

Since $H \cap K = \{e\}$, we have

$$\bar{e} = H, \ \bar{a} = aH, \ \bar{c} = cH, \ \overline{ac} = acH,$$

are distinct elements in the factor group G/H. For example, if $\bar{a} = \bar{c}$, then $aH = cH$. Hence $H = a^{-1}cH$. Thus $ac = a^{-1}c \in H$. This implies $ac \in H \cap K$. Now $ac \neq e$. This contradicts $H \cap K = \{e\}$. We conclude $\bar{a} \neq \bar{c}$.

Now $|G/H| = |G|/|H| = 4$. Thus

$$G/H = \{\bar{e}, \bar{a}, \bar{c}, \overline{ac}\} = \langle \bar{a}, \bar{c} \mid \bar{a}^2 = \bar{c}^2 = (\overline{ac})^2 = \bar{e} \rangle,$$

the Klein 4-group. Thus

$$G = H \cup aH \cup cH \cup acH.$$

This gives:

$$G = \{e, b, b^2, a, ab, ab^2, c, cb, cb^2, ac, acb, acb^2\}.$$

Since H is normal in G, for all $x \in K$, we have $x^{-1}bx \in H$. If $x^{-1}bx = b$ for all $x \in K$, then G would be abelian. This is not so. Hence for some $x \in K$ we have $x^{-1}bx \neq b$. Since any two elements of K_4 generate it with the same generating relations it turns out in this particular case that we get the same G, upto isomorphism, no matter which x we select in K. (If the reader cannot see this, the reader is recommended to work the various cases in detail.) Thus we may suppose that $a^{-1}ba \neq b$. Now $a^{-1}ba \neq e$, otherwise $b = e$. Hence

$$a^{-1}ba = b^2.$$

Thus $\quad ba = ab^2.$

We now want to know bc as an element of G. Because H is normal $c^{-1}bc \in H$. Thus $c^{-1}bc = e$, b or b^2. If $c^{-1}bc = e$, then

$b = e$. This is not so. If $c^{-1}bc = b$, then $(ac)^{-1}b(ac) = c^{-1}a^{-1}bac$ $= c^{-1}b^2c = c^{-1}bc.c^{-1}bc = b^2$, and conversely. If $c^{-}\ bc = b^2$, then $(ac)^{-1}b(ac) = c^{-1}a^{-1}bac = c^{-1}b^2c = c^{-1}bc.c^{-1}bc = b^4 = b$, and conversely. Thus one of c or ac from K_4 transforms b into b and the other transforms b into b^2. Since c and ac may be interchanged in K_4 to give isomorphic groups, it does not matter which we choose to determine G upto isomorphism. (Check if in doubt.) Thus take $c^{-1}bc = b$ and $(ac)^{-1}b(ac) = b^2$.

To sum up we have:

$$G = \{e,\ b,\ b^2,\ a,\ ab,\ ab^2,\ c,\ cb,\ cb^2,\ ac,\ acb,\ acb^2\}\ ,$$

with multiplication determined by:

$a^2 = c^2 = e,\ b^3 = e,\ ac = ca,\ bc = cb,\ ba = ab^2.$

This suffices to give the multiplication table 7.8.3.

To verify that this actually is the multiplication table of a group we produce a known group which has this table as its multiplication table.

Consider D_6 the symmetry group of the regular plane hexagon ABCDEF. Let b be an anti-clockwise rotation through 120 degrees about an axis through O perpendicular to the plane of the hexagon.

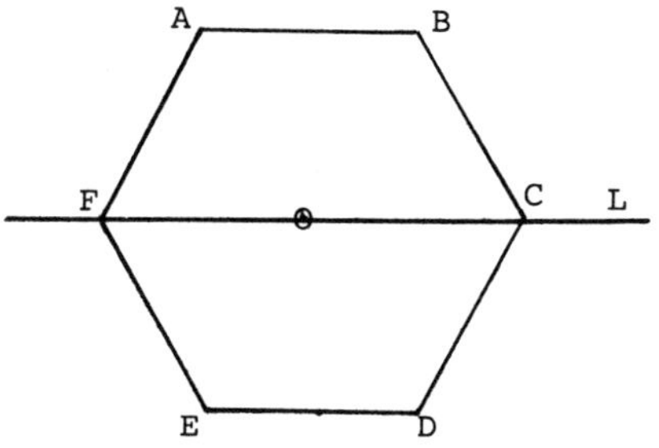

Figure 7.8.3

Let a be a rotation through 180 degrees about the axis L fixed in space.

Let c be an anti-clockwise rotation through 180 degrees about an axis through O perpendicular to the plane of the hexagon.

In terms of a, b, c, where ab means first b then a, D_6 has the distinct elements:

$\{e,\ b,\ b^2,\ a,\ ab,\ ab^2,\ c,\ cb,\ cb^2,\ ac,\ acb,\ acb^2\}$,

and the multiplication table 7.8.3.

	e	b	b^2	a	ab	ab^2	c	cb	cb^2	ac	acb	acb^2
e	e	b	b^2	a	ab	ab^2	c	cb	cb^2	ac	acb	acb^2
b	b	b^2	e	ab^2	a	ab	cb	cb^2	c	acb^2	ac	acb
b^2	b^2	e	b	ab	ab^2	a	cb^2	c	cb	acb	acb^2	ac
a	a	ab	ab^2	e	b	b^2	ac	acb	acb^2	c	cb	cb^2
ab	ab	ab^2	a	b^2	e	b	acb	acb^2	ac	cb^2	c	cb
ab^2	ab^2	a	ab	b	b^2	e	acb^2	ac	acb	cb	cb^2	c
c	c	cb	cb^2	ac	acb	acb^2	e	b	b^2	a	ab	ab^2
cb	cb	cb^2	c	acb^2	ac	acb	b	b^2	e	ab^2	a	ab
cb^2	cb^2	c	cb	acb	acb^2	ac	b^2	e	b	ab	ab^2	a
ac	ac	acb	acb^2	c	cb	cb^2	a	ab	ab^2	e	b	b^2
acb	acb	acb^2	ac	cb^2	c	cb	ab	ab^2	a	b^2	e	b
acb^2	acb^2	ac	acb	cb	cb^2	c	ab^2	a	ab	b	b^2	e

Table 7.8.3

To check this the reader may find it convenient to cut a hexagon out of cardboard, label the vertices, and carry out the various operations on it; thereby incidentally obtaining a permutation representation of D_6 on the vertices of the hexagon.

We conclude that, upto isomorphism, D_6 is the only group of order 12 with one subgroup of order 3, necessarily normal, and three Klein 4-subgroups as its only subgroups of order 4.

The reader is invited to check the following facts about D_6 which follow from the usual considerations. (**Note** a simpler description of D_6 in exercise 7.4.)

Element:	e	b	b^2	a	ab	ab^2	c	cb	cb^2	ac	acb	acb^2
Inverse:	e	b^2	b	a	ab	ab^2	c	cb^2	cb	ac	acb	acb^2
Order:	1	3	3	2	2	2	2	6	6	2	2	2

Conjugacy classes:

$\{e\}$, $\{c\}$, $\{b, b^2\}$, $\{cb, cb^2\}$, $\{a, ab, ab^2\}$, $\{ac, acb, acb^2\}$.

The centre $Z(D_6) = \{e, c\} = \langle c \rangle \cong C_2$.

To find D_6' we argue as follows. $|D_6/H| = 4$, hence D_6/H is abelian. By theorem 4.5.4, $H \supset D_6'$. Thus $|D_6'|$ divides $|H| = 3$. Hence $|D_6'| = 1$ or 3. Now D_6 is not abelian. Hence

$$D_6' \neq \{e\}.$$

Thus $|D_6'| = 3$. Hence

$$D_6' = H = \{e, b, b^2\} = \langle b \rangle \cong C_3 .$$

Subgroup order 1: $\{e\}$.

Subgroups order 2: $H_1 = Z(D_6) = \{e, c\} = \langle c \rangle$,

$H_2 = \{e, a\} = \langle a \rangle$, $H_3 = \{e, ab\} = \langle ab \rangle$, $H_4 = \{e, ab^2\} = \langle ab^2 \rangle$,

$H_5 = \{e, ac\} = \langle ac \rangle$, $H_6 = \{e, acb\} = \langle acb \rangle$,

$H_7 = \{e, acb^2\} = \langle acb^2 \rangle$.

Subgroups order 3: $H_8 = \{e, b, b^2\} = \langle b \rangle = D_6' = H$.

Subgroups order 4: $H_9 = \{e, a, c, ac\} = \langle a, c \rangle = K$,

$H_{10} = \{e, c, ab, acb\} = \langle c, ab \rangle$, $H_{11} = \{e, c, ab^2, acb^2\} = \langle c, ab^2 \rangle$.

Subgroups order 6: $H_{12} = \{e, cb, b^2, c, b, cb^2\} = \langle cb \rangle$,

$H_{13} = \{e, a, ab, ab^2, b, b^2\} = \langle a, b \rangle \cong S_3$,

$H_{14} = \{e, ac, acb, acb^2, b, b^2\} = \langle ac, b \rangle \cong S_3$.

The lattice of subgroups is:

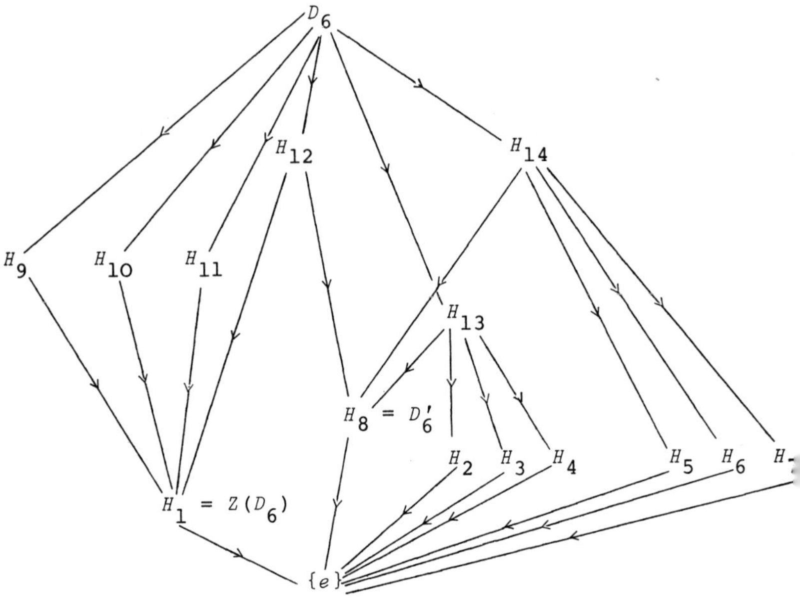

Figure 7.8.4

Factor groups are: $D_6/\{e\} \cong D_6$.

$$D_6/Z(D_6) = \langle \bar{a}, \bar{b}, \bar{c} \rangle,$$

where $\bar{a} = aZ(D_6)$, $\bar{b} = bZ(D_6)$, $\bar{c} = cZ(D_6)$. Now $a^{-1}b^{-1}ab = ab^2ab = b^2 \notin Z(D_6)$. Thus $\bar{a}\bar{b} \neq \bar{b}\bar{a}$. Hence $D_6/Z(D_6)$ is non-abelian of order 6. Thus $D_6/Z(D_6) \cong S_3$. $D_6/D_6' = D_6/H \cong K_4$ as given at the beginning of the discussion on this case (3b).

$$D_6/H_{12} \cong D_6/H_{13} \cong D_6/H_{14} \cong C_2 , \quad D_6/D_6 \cong \{e\} .$$

Also note that $H_1 \cap H_{13} = \{e\}$, and both H_1 and H_{13} are normal in D_6. Moreover $D_6 = H_1H_{13}$. By exercise 5.3, $D_6 \cong H_1 \times H_{13} \cong C_2 \times S_3 \cong C_2 \times D_3$.

7.9 Groups of Order 13

By theorem 3.1.3 all groups of order 13 are cyclic. Thus

$$G = \langle a \mid a^{13} = e \rangle \cong C_{13}.$$

The only subgroups of C_{13} are $\{e\}$ and C_{13} itself.

7.10 Groups of Order 14

Let G be a group of order $14 = 2 \times 7$. By the structure theory for abelian groups in Chapter 5 there is only one abelian group of order 14, namely $C_2 \times C_7 = C_{14}$.

By Sylow's third theorem each non-abelian group of order 14 has just one subgroup H of order 7, necessarily normal, and $1 + 2k \neq 1$ subgroups of order 2, where $1 + 2k$ divides 14. This gives 7 subgroups of order 2. Hence all proper subgroups $\neq \{e\}$ are cyclic of order 2 or 7.

The detailed discussion is exactly parallel to that for non-abelian groups of order 10 in section 7.6. The reader is referred to that section for the details of the argument.

Let $H = \{e, a, a^2, a^3, a^4, a^5, a^6\} = \langle a \mid a^7 = e \rangle$ and let $K = \{e, b\} = \langle b \mid b^2 = e \rangle$. By the usual argument $H \cap K = \{e\}$. Hence $b \notin H$. Thus

$$G/H = \{\bar{e}, \bar{b}\} = \langle \bar{b} \rangle \cong C_2 ;$$

$$\bar{b} = bH = Hb.$$

Hence $$G = H \cup Hb.$$

Thus

$$G = \{e, a, a^2, a^3, a^4, a^5, a^6, b, ab, a^2b, a^3b, a^4b, a^5b, a^6b\} .$$

Now $ba \in bH = Hb$. Thus ba is one of:

$$b, ab, a^2b, a^3b, a^4b, a^5b, a^6b .$$

It turns out as in section 7.6 that we have to take $ba = a^{-1}b$. In this case this gives $ba = a^6b$. Thus

$$G = \{e, a, a^2, a^3, a^4, a^5, a^6, b, ab, a^2b, a^3b, a^4b, a^5b, a^6b\} ,$$

with $\quad a^7 = e, \quad b^2 = e, \quad ba = a^6 b$.

These relations suffice to determine the multiplication table 7.10.1.

	e	a	a^2	a^3	a^4	a^5	a^6	b	ab	a^2b	a^3b	a^4b	a^5b	a^6b
e	e	a	a^2	a^3	a^4	a^5	a^6	b	ab	a^2b	a^3b	a^4b	a^5b	a^6b
a	a	a^2	a^3	a^4	a^5	a^6	e	ab	a^2b	a^3b	a^4b	a^5b	a^6b	b
a^2	a^2	a^3	a^4	a^5	a^6	e	a	a^2b	a^3b	a^4b	a^5b	a^6b	b	ab
a^3	a^3	a^4	a^5	a^6	e	a	a^2	a^3b	a^4b	a^5b	a^6b	b	ab	a^2b
a^4	a^4	a^5	a^6	e	a	a^2	a^3	a^4b	a^5b	a^6b	b	ab	a^2b	a^3b
a^5	a^5	a^6	e	a	a^2	a^3	a^4	a^5b	a^6b	b	ab	a^2b	a^3b	a^4b
a^6	a^6	e	a	a^2	a^3	a^4	a^5	a^6b	b	ab	a^2b	a^3b	a^4b	a^5b
b	b	a^6b	a^5b	a^4b	a^3b	a^2b	ab	e	a^6	a^5	a^4	a^3	a^2	a
ab	ab	b	a^6b	a^5b	a^4b	a^3b	a^2b	a	e	a^6	a^5	a^4	a^3	a^2
a^2b	a^2b	ab	b	a^6b	a^5b	a^4b	a^3b	a^2	a	e	a^6	a^5	a^4	a^3
a^3b	a^3b	a^2b	ab	b	a^6b	a^5b	a^4b	a^3	a^2	a	e	a^6	a^5	a^4
a^4b	a^4b	a^3b	a^2b	ab	b	a^6b	a^5b	a^4	a^3	a^2	a	e	a^6	a^5
a^5b	a^5b	a^4b	a^3b	a^2b	ab	b	a^6b	a^5	a^4	a^3	a^2	a	e	a^6
a^6b	a^6b	a^5b	a^4b	a^3b	a^2b	ab	b	a^6	a^5	a^4	a^3	a^2	a	e

Table 7.10.1

To verify that this actually is the multiplication table of a group we consider D_7 the symmetry group of a regular plane polygon ABCDEFG with 7 sides, i.e. a regular heptagon.

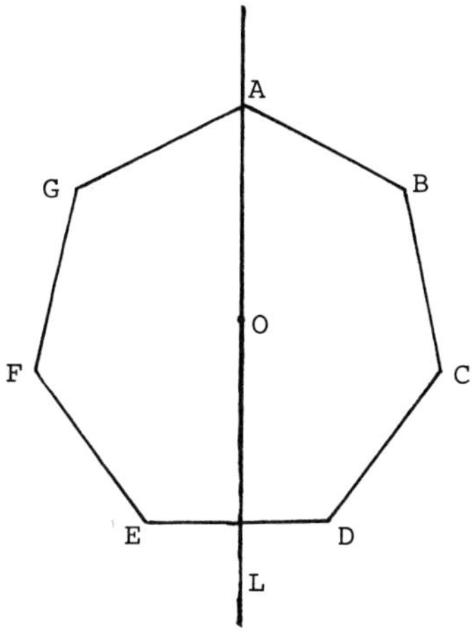

Figure 7.10.1

Let L through the centre O be fixed in space and perpendicular to a side of the heptagon. Let a be a rotation through 360/7 degrees in an anti-clockwise sense about an axis through O perpendicular to the plane of the heptagon.

Let b be a rotation about the line L through 180 degrees.

In terms of a and b the symmetry group D_7 has the multiplication table 7.10.1. Remember that yx means: first x then y.

It follows that the dihedral group D_7 is the only non-abelian group of order 14.

In the usual ways we obtain the following information about D_7.

Element:	e	a	a^2	a^3	a^4	a^5	a^6	b	ab	a^2b	a^3b	a^4b	a^5b	a^6b
Inverse:	e	a^6	a^5	a^4	a^3	a^2	a	b	ab	a^2b	a^3b	a^4b	a^5b	a^6b
Order:	1	7	7	7	7	7	7	2	2	2	2	2	2	2

Conjugacy classes:

$\{e\}$, $\{a, a^6\}$, $\{a^2, a^5\}$, $\{a^3, a^4\}$,

$\{b, ab, a^2b, a^3b, a^4b, a^5b, a^6b\}$.

The centre $Z(D_7)$ is: $\{e\}$.

The commutator subgroup is:

$$D_7' = \{e, a, a^2, a^3, a^4, a^5, a^6\} = \langle a \rangle = H.$$

Subgroup order 1: $\{e\}$.

Subgroups order 2: $H_1 = \{e, b\} = \langle b \rangle$, $H_2 = \{e, ab\} = \langle ab \rangle$,

$H_3 = \{e, a^2b\} = \langle a^2b \rangle$, $H_4 = \{e, a^3b\} = \langle a^3b \rangle$,

$H_5 = \{e, a^4b\} = \langle a^4b \rangle$, $H_6 = \{e, a^5b\} = \langle a^5b \rangle$,

$H_7 = \{e, a^6b\} = \langle a^6b \rangle$.

Subgroups order 7: $H_8 = D_7' = \{e, a, a^2, a^3, a^4, a^5, a^6\} = \langle a \rangle = H$.

The only subgroup order 14 is of course D_7 itself. The lattice of subgroups is:

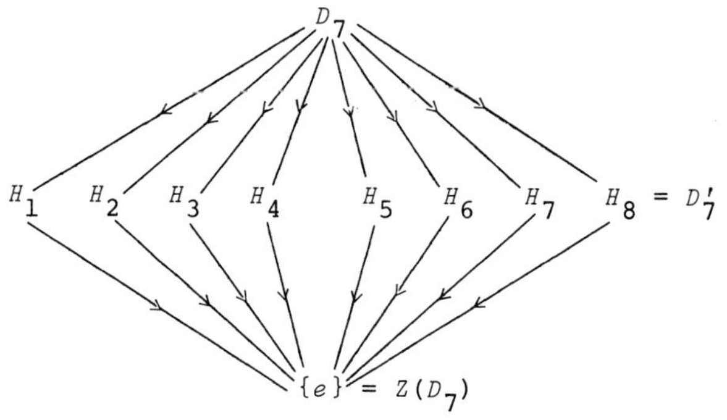

Figure 7.10.2

Factor groups are: $D_7/\{e\} \cong D_7$, $D_7/D_7' \cong C_2$, and $D_7/D_7 \cong \{e\}$.

7.11 Groups of Order 15

Let G be a group of order 15. We have $15 = 3 \times 5$. Moreover $3 \nmid (5 - 1)$. By theorem 6.4.1 G is cyclic. Hence

$$G = C_3 \times C_5 = C_{15},$$

$$Z(C_{15}) = C_{15}$$

and

$$C_{15}' = \{e\} .$$

All subgroups are normal.

The lattice of subgroups is:

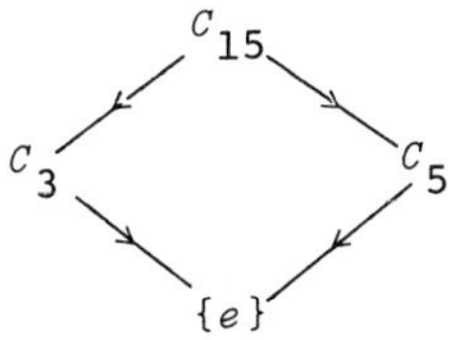

Figure 7.11.1

Factor groups are: $C_{15}/\{e\} \cong C_{15}$, $C_{15}/C_3 \cong C_5$, $C_{15}/C_5 \cong C_3$, $C_{15}/C_{15} \cong \{e\} = C_1$.

7.12 Summary

We summarise a few results on the groups of orders 1 to 15 in the following table.

Order	Abelian Number	Abelian Type	Non-abelian Number	Non-abelian Type
1	1	C_1	0	-
2	1	C_2	0	-
3	1	C_3	0	-
4	2	C_4, K_4 (Klein)	0	-
5	1	C_5	0	-
6	1	$C_6 = C_2 \times C_3$	1	$S_3 = D_3$
7	1	C_7	0	-
8	3	C_8, $C_2 \times C_4$, $C_2 \times C_2 \times C_2$	2	D_4, Q
9	2	C_9, $C_3 \times C_3$	0	-
10	1	$C_{10} = C_2 \times C_5$	1	D_5
11	1	C_{11}	0	-
12	2	$C_{12} = C_3 \times C_4$, $C_2 \times C_2 \times C_3 = K_4 \times C_3$	3	A_4, M, $D_6 = C_2 \times D_3$
13	1	C_{13}	0	-
14	1	$C_{14} = C_2 \times C_7$	1	D_7
15	1	$C_{15} = C_3 \times C_5$	0	-

Table 7.12.1

Note the close similarity between the groups of orders 6, 10 and 14. This carries through to all groups of order $2p$, where p is an odd prime.

Note that some of the work of this chapter can be streamlined by use of the theory of presentation of groups. As mentioned before, 'Presentation of Groups' — Johnson (C.U.P.) 1976 provides a good introduction to this theory. More advanced references are given in the recommendations for further reading at the end of this book.

The ideas discussed in section 4.6 can also be used to shorten the work of this chapter. The reader is invited to explore this possibility.

EXERCISES 7

Verify the following assertions: 1 to 6.

1. $S_3 \cong \langle a, b \mid a^3 = b^2 = (ab)^2 = e \rangle$.
2. $D_4 \cong \langle x, y \mid x^4 = y^2 = (xy)^2 = e \rangle$.
3. $D_5 \cong \langle x, y \mid x^5 = y^2 = (xy)^2 = e \rangle$.
4. $D_6 \cong \langle x, y \mid x^6 = y^2 = (xy)^2 = e \rangle$.
5. $A_4 \cong \langle s, t \mid s^3 = t^2 = (st)^3 = e \rangle$.
6. $D_7 \cong \langle a, b \mid a^7 = b^2 = (ab)^2 = e \rangle$.
7. A finite group G is said to be ZS-metacylic (see 'Generators and Relations for Discrete Groups' by Coxeter and Moser published by Springer, 1972) if G' is cyclic, G/G' is cyclic, and **all** its Sylow subgroups are cyclic.

   Verify that D_5, M, D_7, are ZS-metacylic.

CHAPTER 8.
EPILOGUE

8.1 Introduction

A list of all non-abelian groups of orders upto 30 is given in Table 1 on page 134 of 'Generators and Relations for Discrete Groups' — Coxeter and Moser (Springer, 1972). Of course the abelian groups are easily dealt with using the structure theory of Chapter 5. Elementary methods of obtaining low-order groups along similar lines to those used in Chapter 7 are given in sections 126 and 127 of the classic 'Theory of Groups of Finite Order' by Burnside, 1911 (republished by Dover, 1955). In sections 112 to 118 of the same book elementary methods are given for determining all groups of orders p^2, p^3, and p^4 for a prime p.

All possible groups of order 2^n, $n \leq 6$, and their interconnections are discussed and tabulated in a monumental work 'The Groups of Order 2^n ($n \leq 6$)' by M. Hall and J. K. Senior (Macmillan, 1964). One of the difficulties of a work such as this is to keep track of the different groups that arise so that repetitions are avoided. To do this, Marshall Hall and Senior use a classification based on the idea of isoclinic groups due to Philip Hall and discussed in his paper 'The Classification of Prime-Power Groups' (Journal für die reine.u.ang.Math.—vol. 182, 1940, pp. 130-141).

8.2 Construction of Finite Groups

A general method for constructing all finite groups arises from the following considerations.

Given a group G and a normal subgroup N we have seen how to construct the factor or quotient group G/N. It is a natural converse step to consider the following problem. Given a group N and a group H can we find a group G such that N is a normal subgroup of G and $G/N \cong H$.

This is the **extension** problem. G is said to be an **extension** of N by H. The theory of such constructions has been extensively studied since Schreier first considered the problem in the 1920's. It is now fairly well understood and has been absorbed into a wider theory, with considerable ramifications, called the cohomology theory of groups. (See 'Theory of Groups', Vol. 2 —Kurosh; 'Cohomological Topics in Group Theory' —Gruenberg; and 'Homology' —MacLane.)

Given any finite group G we can always construct, at least in theory, a series of groups of the form

$$\{e\} \subset H_1 \subset H_2 \subset \cdots \subset H_n = G \quad ,$$

where each H_i is a **normal** subgroup of H_{i+1} and where each factor group H_{i+1}/H_i is **simple**. (See definition 4.2.1.) The latter is the same as saying that each H_i is a **maximal** normal subgroup of H_{i+1}. This follows from theorem 4.7.1. Thus in a finite group G we can start with $\{e\} \subset G$ and then if G is not already simple we can insert a normal subgroup N of G to get $\{e\} \subset N \subset G$. If N is not simple we can get the series $\{e\} \subset N' \subset N \subset G$. If G/N is not simple we can get $\{e\} \subset N' \subset N \subset N'' \subset G$. We repeat the above. Because G is finite, this process must terminate after a finite number of steps with each factor of the series (i.e. each factor group such as N/N', etc.) a simple group. Such a series is called a **composition** series.

Our argument above shows that any finite group has a composition series. If we combine this with the extension theory mentioned before we have a general method of constructing all finite groups.

We extend the identity group $\{e\}$ by a simple group to get G_1, say, and then we extend G_1 by a simple group to get G_2, say, and then we repeat the process any number of times. Thus the problem of the structure of the general finite group can be broken down into two problems:

(1) the extension problem,

(2) the discovery of all finite simple groups.

These are major preoccupations of workers in the field of finite groups at the moment. An excellent study of most of the known simple groups is given in 'Simple Groups of Lie Type' — Carter (Wiley, 1972). As we have proved in theorem 4.2.2 the alternative groups A_n, $n \geq 5$, are simple. They provide us with an example of an infinite sequence of simple groups. Some discussion of the current attempts to verify that the known simple groups are all the simple groups that can exist is given in: 'Finite Simple Groups' —Powell and Higman (Academic Press, 1971). Progress since then suggests that this programme may soon be completed. The work involves a considerable amount of technical apparatus. One of the chief tools is the theory of group representations and characters. A good introduction to this theory is 'Introduction to Group Characters' by Ledermann (C.U.P., 1977). An excellent advanced text is: 'Group Representation Theory A and B' by Dornhoff (Marcel Dekker, 1971 and 1972).

For **infinite** groups we can still consider series of the form $\{e\} \subset H_1 \subset H_2 \subset \cdots \subset H_n = G$, where each H_i is a normal subgroup of H_{i+1}. However, it is not always possible to make each factor of the series H_{i+1}/H_i simple. This type of series is called a **subnormal** series. **Note** that there are several other terms used in the literature. One such is normal series. We reserve this for the case when all the subgroups are normal in G.

The **length** of the series is the **number** of **factors** in the series, i.e. one less than the number of groups. A **refinement** of the series is obtained by inserting extra groups. For example $\{e\} \subset H_1 \subset H_2 \subset G$ is a refinement of $\{e\} \subset H_1 \subset G$. Of course we must **always** have H_i **normal** in H_{i+1}.

Two subnormal series of a group G are said to be **isomorphic** if there is a one-one correspondence between their factors such that corresponding factors are isomorphic as groups. We give these definitions for use in a moment.

8.3 Solvable and Nilpotent Groups

Rather than study groups of specific orders it has proved useful to define a class of groups by some property they have in common and to study the classification and structure of this restricted class of groups.

A good example of this is provided by the theory of finitely generated abelian groups given in Chapter 5. Such groups are direct products of cyclic groups of prime power order. In the finite case, by allowing a direct product of arbitrary groups of prime power order, we get a natural generalisation to the class of **nilpotent** groups. As shown in exercise 6.7 this is the class of groups whose Sylow subgroups are all normal. There is one for each prime divisor of the order of the group.

For infinite groups a group G is said to be **nilpotent** if it has a **central** series; that is a **normal** series, as discussed at the end of section 8.2; $\{e\} = G_0 \subset G_1 \subset G_2 \subset \cdots \subset G_n = G$, in which the factor G_{i+1}/G_i lies in the **centre** of G/G_i for $i = 0, 1, 2, 3, 4, 5, \ldots, (n - 1)$. This is equivalent to our previous definition for finite groups. The proof of this will not be given here.

An important generalisation of the idea of a nilpotent group is that of a **solvable** group. This is a group with a subnormal series: $\{e\} = G_0 \subset G_1 \subset G_2 \subset \cdots \subset G_n = G$, in which **all** factors G_{i+1}/G_i are **abelian**. In particular all abelian groups are solvable.

Solvable groups are important in Galois theory. They are closely associated with the solution of equations as mentioned in the introduction section 1.1. Roughly speaking solvable groups are associated, as permutation groups on the roots of the equation, with equations that are solvable in terms of successive extractions of roots. A good introduction to Galois theory is: 'Galois Theory' —Stewart (Chapman and Hall, 1973).

In order to make this book more useful to those whose main interest is in Galois theory we give brief statements and proofs of some of the most basic results on solvable groups. First we need some preliminary results of interest in themselves.

8.4 The Isomorphism Theorems

First a useful result:

THEOREM 8.4.1. *Let f be a homomorphism from G into L $f\colon\ G \to L$. Let f be restricted to a subgroup H of G to give the homomorphism $f_r\colon\ H \to L$ defined formally by $f_r(h) = f(h)$ for all $h \in H$. Then the kernel of f_r is $\operatorname{Ker} f \cap H$. Thus $\operatorname{Ker} f_r = \operatorname{Ker} f \cap H$.*

PROOF Let $h \in \operatorname{Ker} f_r$. Then $f_r(h) = e$. Thus $h \in \operatorname{Ker} f$. Now $h \in H$. Thus $h \in \operatorname{Ker} f \cap H$.

Conversely if $h \in \operatorname{Ker} f \cap H$, then $f(h) = e$. Since $h \in H$, $f_r(h)$ is defined and equals $f(h) = e$. Thus $h \in \operatorname{Ker} f_r$. Hence

$$\operatorname{Ker} f_r = \operatorname{Ker} f \cap H.$$

We give the next theorem in a slightly more general form than usual.

THEOREM 8.4.2. *Let H be any subgroup of a group G. Let N be any **normal** subgroup of G. Let M be any **normal** subgroup of H. Then*

$$HN/MN \simeq H/M(H \cap N) .$$

PROOF We first show that MN is normal in HN. Since N is normal in G, $HN = NH$, so that HN is certainly a subgroup of G. Similarly, MN is a subgroup of G which is contained in HN since $M \subset H$. Thus MN is a subgroup of HN. Take $x \in MN$ and $y \in HN$. We have to show that $y^{-1}xy \in MN$. We can write $x = mn_1$ and $y = hn_2$ for certain elements $m \in M$, $h \in H$, n_1, $n_2 \in N$. Thus

$$y^{-1}xy = n_2^{-1}h^{-1}mn_1hn_2 = n_2^{-1}h^{-1}mhn_3n_2,$$

where $hn_3 = n_1h$ since $hN = Nh$ as N is normal in G. Now M is normal in H. Hence $h^{-1}mh \in M$. Thus we can write

$$h^{-1}mh = m_1$$

for some $m_1 \in M$. Thus

$$y^{-1}xy = n_2^{-1}m_1n_3n_2 \ .$$

Now N is normal in G, thus $Nm_1 = m_1N$. In particular $n_2^{-1}m_1 = m_1n_4$, for some $n_4 \in N$. Hence

$$y^{-1}xy = m_1(n_4n_3n_2) \in MN$$

as required.

Now consider the natural homomorphism f_{MN} from HN onto HN/MN. We restrict f_{MN} to the subgroup H of HN. Call this restriction f'_{MN}. We note that f'_{MN} is a homomorphism of H onto HN/MN. To show this let y be any element of HN/MN. We may write $y = hn(MN)$, where hn is an element of HN which represents the coset $hn(MN)$ containing hn. Now $MN = NM$, since N is normal in G. Thus

$$hn(MN) = hn(NM) = hNM = hMN.$$

Thus $y = hMN$. But then

$$f'_{MN}(h) = f_{MN}(h) = hMN = y \ .$$

Thus f'_{MN} is an epimorphism from H onto HN/MN. By theorem 4.5.3 we have $HN/MN \cong H/\text{Ker}\, f'_{MN}$. Then using theorem 8.4.1 we have

$$\text{Ker}\, f'_{MN} = (\text{Ker}\ f_{MN}) \cap H \ .$$

Now $\text{Ker}\, f_{MN} = MN$. Thus

$$\text{Ker}\, f'_{MN} = (MN) \cap H \ .$$

Finally we show that $(MN) \cap H = M(H \cap N)$. Let $x \in (MN) \cap H$. Then $x = mn = h$; $m \in M$, $n \in N$, $h \in H$. Thus $n = m^{-1}h \in H$, because $M \subset H$. Thus $n \in H \cap N$. Hence

$$x = mn \in M(H \cap N).$$

Thus $\qquad (MN) \cap H \subset M(H \cap N).$

Now suppose that $x \in M(H \cap N)$. Then $x = mn$, where $m \in M$ and $n \in H \cap N$. Thus m and n are in H. Hence $mn \in H$. Now $mn \in MN$. Thus

$$x = mn \in (MN) \cap H \ .$$

Hence $\qquad M(H \cap N) \subset (MN) \cap H.$

The two inclusions together give:

$$(MN) \cap H = M(H \cap N) .$$

Thus $\quad \operatorname{Ker} f'_{MN} = M(H \cap N)$.

Putting this result together with the isomorphism established earlier we get:

$$HN/MN \cong H/M(H \cap N),$$

as required.

The usual form of the previous theorem is:

THEOREM 8.4.3. *If H is any subgroup of a group G and N is any* ***normal*** *subgroup of G, then HN is a subgroup of G and we have the isomorphism:*

$$HN/N \cong H/H \cap N .$$

PROOF In the previous theorem take $M = \{e\}$. Then $HN/MN \cong H/M(H \cap N)$ reduces at once to

$$HN/N \cong H/H \cap N ,$$

as required.

For a direct proof of this result see exercise 8.1.

The following result is a rather technical theorem often called Zassenhaus's Lemma. Our theorem 8.4.2 provides exactly what is required for a proof.

THEOREM 8.4.4. *If U and V are subgroups of a group G and u and v are normal subgroups of U and V respectively, then the following three factor groups are isomorphic:*

$$u(U \cap V)/u(U \cap v) \cong U \cap V/(u \cap V)(v \cap U) \cong v(U \cap V)/v(u \cap V) .$$

PROOF Take $G = U$, $H = U \cap V$, $M = U \cap v$, $N = u$, in the statement of theorem 8.4.2.

We leave the reader to check that $U \cap v$ is normal in $U \cap V$. Then the conditions of the theorem hold. We deduce that

$$u(U \cap V)/u(U \cap v) \cong U \cap V/(U \cap v)(U \cap V \cap u) = U \cap V/(U \cap v)(u \cap V).$$

A similar result is obtained taking:

$$G = V,\ H = U \cap V,\ M = u \cap V,\ N = v .$$

We get:

$$v(U \cap V)/v(u \cap V) \cong U \cap V/(u \cap V)(v \cap U) \ .$$

Since $U \cap v$ is normal in $U \cap V$, we have

$$(U \cap v)(u \cap V) = (u \cap V)(U \cap v).$$

Putting the previous isomorphisms together we get the required theorem.

8.5 The Schreier — Jordan — Hölder Theorem

This theorem has a rather long name because it is called after the three mathematicians who each contributed to the complete result as it is now given. We call it prosaically:

THEOREM 8.5.1. *Any two subnormal series of an arbitrary group G have isomorphic refinements. These can be given without repetitions if the original series are without repetitions.*

PROOF Suppose the group G has the two subnormal series:

(1) $G = A_0 \supset A_1 \supset A_2 \supset \cdots \supset A_k = \{e\}$,

(2) $G = B_0 \supset B_1 \supset B_2 \supset \cdots \supset B_\ell = \{e\}$.

Note that we are using a different numbering from that used in section 8.2.

Set $A_{ij} = A_i(A_{i-1} \cap B_j)$ for $i = 1, 2, 3, 4, \ldots, k$ and $j = 0, 1, 2, 3, \ldots, \ell$.

Set $B_{ji} = B_j(B_{j-1} \cap A_i)$ for $i = 0, 1, 2, 3, \ldots, k$ and $j = 1, 2, 3, 4, \ldots, \ell$.

We now use theorem 8.4.4 (Zassenhaus's Lemma) with the following identifications:

$$U = A_{i-1}, \quad u = A_i, \quad V = B_{j-1}, \quad v = B_j \ .$$

From the Lemma we have:

$$u(U \cap V)/u(U \cap v) \cong v(U \cap V)/v(u \cap V).$$

Hence $\quad A_i(A_{i-1} \cap B_{j-1})/A_i(A_{i-1} \cap B_j) \cong B_j(A_{i-1} \cap B_{j-1})/B_j(A_i \cap B$

Thus we have

$$(3): A_{i(j-1)}/A_{ij} \cong B_{j(i-1)}/B_{ji} .$$

It now follows that A_{ij} and B_{ji} are subgroups of G with the inclusions:

$$A_{i-1} = A_{i0} \supset A_{i(j-1)} \supset A_{ij} \supset A_{i\ell} = A_i ,$$

$$B_{j-1} = B_{j0} \supset B_{j(i-1)} \supset B_{ji} \supset B_{jk} = B_j ,$$

where A_{ij} is normal in $A_{i(j-1)}$ and B_{ji} is normal in $B_{j(i-1)}$.

If all the A_{ij} $(j = 1, 2, 3, \ldots, (\ell - 1))$ are inserted between A_{i-1} and A_i $(i = 1, 2, \ldots, k)$ in (1), and all the B_{ji} $(i = 1, 2, 3, \ldots, (k - 1))$ are inserted between B_{j-1} and B_j $(j = 1, 2, 3, \ldots, \ell)$ in (2), we get isomorphic refinements of (1) and (2) because of (3).

Even if (1) and (2) have no repetitions, i.e. all the inclusions are **proper**, $A_i \neq A_{i-1}$, $B_j \neq B_{j-1}$; it is possible for the refinements that we have obtained to have repetitions. However, these repetitions can be deleted without affecting the isomorphism of the refined series. This can be seen as follows.

If $A_{i(j-1)} = A_{ij}$ then $A_{i(j-1)}/A_{ij} \cong \{e\}$. Now $A_{i(j-1)}/A_{ij} \cong B_{j(i-1)}/B_{ji}$. Hence

$$B_{j(i-1)}/B_{ji} \cong \{e\}.$$

Thus $\quad B_{j(i-1)} = B_{ji}$.

Hence if we strike out $A_{i(j-1)}$ then we can also strike out $B_{j(i-1)}$, and the remaining isomorphisms of corresponding factors are not affected.

The theorem follows.

8.6 Some Basic Results on Solvable Groups

The following celebrated result assures us that there are a lot of solvable groups about.

THEOREM 8.6.1. *Every finite group of odd order is solvable.*

PROOF The monumental proof of this theorem by W. Feit and J. G. Thompson in 1963 covers pages 755 to 1029 of the *Pacific Journal of Mathematics* **13**.

In this proof many techniques were introduced which have proved useful in the work on finite simple groups mentioned in section 8.2.

THEOREM 8.6.2. *Let G be a solvable group. Then any subnormal series of G can be refined until it is a subnormal series with abelian factors. (Such a series is called a* ***solvable series***.*)*

PROOF Let $G = G_0 \supset G_1 \supset \cdots \supset G_r = \{e\}$ be a subnormal series of a solvable group G. Since G is solvable it has a solvable series, say, $G = A_0 \supset A_1 \supset \cdots \supset A_s = \{e\}$.

By theorem 8.5.1 we can refine both series until they are isomorphic. If we can show that the solvable series remains a solvable series after refinement, then the theorem will follow.

Suppose $A_i \supset A_{i+1}$ occurs in the solvable series with A_i/A_{i+1} abelian. Consider $A_i \supset A \supset A_{i+1}$. We have to show that A_i/A and A/A_{i+1} are both abelian. Now $A_i/A \cong (A_i/A_{i+1})/(A/A_{i+1})$ by theorem 4.8.1. However, A/A_{i+1} is a subgroup of A_i/A_{i+1} and therefore is abelian. From the above isomorphism A_i/A is abelian. Here we have used the fact that factor groups of abelian groups are also abelian.

The theorem follows.

THEOREM 8.6.3. *If G is a finite solvable group then G possesses a composition series in which all factors are cyclic groups of prime order.*

PROOF Refine any solvable series until it is a composition series. As shown in the proof of the previous theorem the series is still a solvable series. Hence all factors are simple abelian groups. By exercise 4.20 all such groups are cyclic of prime order. The theorem follows.

We are now going to find a useful criterion for a group to be solvable. First we give a preliminary theorem and a definition.

THEOREM 8.6.4. *Let G be a group. Let $G = G^{(0)}$. Let $G^{(i+1)}$ denote the derived or commutator group of $G^{(i)}$. In our previous notation $G^{(1)} = G'$. Then $G^{(i)}$ is a* ***normal*** *subgroup of G, for all i.*

PROOF True for $i = 0$ and 1; in the latter case by theorem 4.5.4. Suppose $H = G^{(i)}$ is normal in G. Consider

$$G^{(i+1)} = H' = \langle \{a^{-1}b^{-1}ab \mid a, b \in H\} \rangle .$$

Let $g \in G$. Let $x \in H'$. To show H' normal in G we have to show that $g^{-1}xg \in H'$.

Now $x = y_1 y_2 \ldots y_t$, where $y_i \in \{a^{-1}b^{-1}ab \mid a, b \in H\}$; $i = 1, 2, \ldots, t$. Thus

$$g^{-1}xg = g^{-1}y_1g.g^{-1}y_2g. \ \ldots \ .g^{-1}y_tg .$$

Hence we have to show that $g^{-1}y_ig \in H'$ for $i = 1, 2, \ldots, t$. Now $y_i = a^{-1}b^{-1}ab$, for some $a, b \in H$. Thus

$$g^{-1}y_ig = (g^{-1}ag)^{-1}(g^{-1}bg)^{-1}(g^{-1}ag)(g^{-1}bg) .$$

Now H is normal in G. Thus $h_1 = g^{-1}ag$ and $h_2 = g^{-1}bg$ are elements of H. Thus

$$g^{-1}y_ig = h_1^{-1}h_2^{-1}h_1h_2 = [h_1, h_2]; \quad h_1, h_2 \in H .$$

Hence $\quad g^{-1}y_ig \in H'$.

Thus $H' = G^{(i+1)}$ is normal in G if $G^{(i)}$ is normal in G. The theorem follows by induction on i.

DEFINITION 8.6.1. *Let G be a group. The normal (by theorem 8.6.4) series:*

$$G = G^{(0)} \supset G^{(1)} \supset G^{(2)} \supset G^{(3)} \supset \ldots \supset G^{(i)} \supset \ldots$$

is called the ***derived series*** *of G.*

The criterion for a group to be solvable mentioned earlier is:

THEOREM 8.6.5. *Let G be a group. Then G is solvable if and only if the derived series terminates in $G^{(k)} = \{e\}$ for some finite k.*

PROOF If $G^{(k)} = \{e\}$ for some finite k, then the derived series is a solvable series because $G^{(i)}/G^{(i+1)}$ is abelian. Hence G is solvable.

Conversely if G is solvable it has a solvable series (see theorem 8.6.2). Let $G = A_0 \supset A_1 \supset A_2 \supset \dots \supset A_s = \{e\}$

be a solvable series for G. Then G/A_1 is abelian. By theorem 4.5.4 $A_1 \supset G^{(1)}$. Suppose $A_i \supset G^{(i)}$. Now A_i/A_{i+1} is abelian. By theorem 4.5.4 $A_{i+1} \supset A_i'$. Now $A_i \supset G^{(i)}$. Thus

$$A_i' \supset (G^{(i)})' = G^{(i+1)}.$$

Hence $$A_{i+1} \supset G^{(i+1)}.$$

By a finite induction on i, eventually we have:

$$\{e\} = A_s \supset G^{(s)} .$$

Thus $$G^{(s)} = \{e\} .$$

The theorem follows.

It is useful to know that solvability is inherited by subgroups and factor groups. Formally we have:

THEOREM 8.6.6. *All subgroups of a solvable group G are themselves solvable.*

PROOF Let $G = A_0 \supset A_1 \supset A_2 \supset \dots \supset A_s = \{e\}$ be a solvable series for G. Let H be any subgroup of G. Put $H_i = H \cap A_i$. This gives the following series:

$$H = H_0 \supset H_1 \supset H_2 \supset \dots \supset H_s = \{e\}.$$

We leave the reader to verify that H_{i+1} is normal in H_i. Now

$$H_{i+1} = H \cap A_{i+1} = H \cap A_i \cap A_{i+1} = H_i \cap A_{i+1} .$$

Thus $$H_i/H_{i+1} = H_i/H_i \cap A_{i+1} \cong H_iA_{i+1}/A_{i+1} \text{ by theorem 8.4.3.}$$

Note that in the application of this theorem the role of G is played by A_i with A_{i+1} normal in A_i and $H_i = H \cap A_i$ a subgroup of A_i.

Now $H_iA_{i+1}/A_{i+1} \subset A_i/A_{i+1}$ which is abelian. Hence H_i/H_{i+1} is abelian. Thus

$$H = H_0 \supset H_1 \supset H_2 \supset H_3 \supset \dots \supset H_s = \{e\}$$

is a solvable series. Hence H is a solvable group.

THEOREM 8.6.7. *All epimorphic images (equivalently all factor groups by theorem 4.5.3) of a solvable group are themselves solvable.*

PROOF Let $G = A_0 \supset A_1 \supset \cdots \supset A_s = \{e\}$ be a solvable series for G. Let f be an epimorphism of G onto H. Let $H = f(G)$. Consider the following series for H:

$$H = f(A_0) \supset f(A_1) \supset \cdots \supset f(A_s) = \{e\} .$$

By exercise 4.30 this is a subnormal series for H. Define a function $f_i\colon A_i/A_{i+1} \to f(A_i)/f(A_{i+1})$ by $f_i(\bar{x}) = \overline{f(x)}$, where $\bar{x}$ denotes the coset xA_{i+1} and $\overline{f(x)}$ denotes the coset $f(x)f(A_{i+1})$ considered as elements of the respective factor groups. We leave the reader to verify that if $\bar{x} = \bar{y}$ then $\overline{f(x)} = \overline{f(y)}$. Hence f_i does **not** depend upon the choice of x from the coset xA_{i+1}. Since every element in $f(A_i)$ is of the form $f(x)$ for some $x \in A_i$, f_i must be onto.

Finally

$$f_i(\bar{x}.\bar{w}) = f_i(\overline{xw}) = \overline{f(xw)} = \overline{f(x)f(w)} = \overline{f(x)}.\ \overline{f(w)} = f_i(\bar{x})f_i(\bar{w}) .$$

Thus f_i is an epimorphism of A_i/A_{i+1} onto $f(A_i)/f(A_{i+1})$.

Now A_i/A_{i+1} is abelian. Hence $f(A_i)/f(A_{i+1})$ is abelian. Thus

$$H = f(A_0) \supset f(A_1) \supset f(A_2) \supset \cdots \supset f(A_s) = \{e\}$$

is a solvable series for H. Hence H is a solvable group.

Note: A neat alternative proof of theorem 8.6.7 goes as follows.

Let G/N be any factor group of G. By theorem 8.6.2 refine the subnormal series $G \supset N \supset \{e\}$ until it becomes a solvable series:

$$G = G_0 \supset G_1 \supset G_2 \supset \cdots \supset N \supset \cdots \supset \{e\} .$$

Then theorem 4.8.1 assures us that

$$G/N = G_0/N \supset G_1/N \supset \cdots \supset N/N = \{\bar{e}\}$$

is a solvable series for G/N, where here $\{\bar{e}\}$ is the identity subgroup of G/N. Thus G/N is a solvable group.

We make some final observations. By theorems 8.6.1 and 8.6.3, we can construct all finite groups of **odd** order by the method discussed in section 8.2 using only cyclic groups of order p, for various primes p, as the simple groups in the construction.

A famous theorem of Burnside asserts that all groups of order $p^a.q^b$, p and q primes, are solvable. This, together with theorem 8.6.3, shows that all groups of order p^aq^b can also be constructed as above.

This method of construction has been exploited in order to find all possible groups of a given order n when the order n can be expressed in some specified way, such as $n = p^2q$, p and q primes.

For further details the reader is advised to consult [6] section 59 and the references given in the footnote on page 80.

EXERCISES 8

1. Prove theorem 8.4.3 (that $HN/N \cong H/H \cap N$) directly.

2. Let N be normal in G. Let P be a Sylow p-subgroup of G. Prove that PN/N is a Sylow p-subgroup of G/N.

3. Let m be a positive integer. Let T denote a subgroup of S_n containing m odd permutations. Use theorem 8.4.3 to prove that T has a normal subgroup of index 2. Deduce that T has order $2m$ and hence T has the same number of even permutations as odd. (Compare with exercises 3.11 and 4.15.)

4. Let G be a finite group. Let $H \neq G$ be a subgroup of G such that $|G|$ does **not** divide $[G : H]!$. Prove that H must contain a normal subgroup of G other than $\{e\}$.

5. Assume that A_5 is simple. Prove that A_5 contains no subgroup of order greater than 12 other than A_5 itself. (This simplifies the task of finding all subgroups of A_5.)

6. Prove that any two composition series of the **same** group are isomorphic.

7. Let G be any group not necessarily finite. Let G have a composition series. Prove that every subnormal series can be refined to a composition series. Also prove that every **normal** subgroup of G can be included in some composition series.

8. Give an example of a group which does **not** have a composition series.

9. Show that the symmetric groups S_3 and S_4 are solvable groups.

10. Show that the symmetric group S_n is **not** solvable for $n \geq 5$.

MISCELLANEOUS EXERCISES

1. With the help of the class equation (theorem 4.4.2) or otherwise, find all finite groups G which satisfy the condition that $N_G(a)$ the normaliser of a in G equals $\{e, a, a^{-1}\}$ for all $a \in G$, $a \neq e$. Here we allow a to equal a^{-1} for some $a \in G$.

2. Let f be a homomorphism of a group G into a group H. Let a be an element of G. Show that the order of $f(a)$ in H divides the order of a in G. Let K be a normal subgroup of index n in the group G. Let H be a subgroup of G with $|H| = m$ coprime with n. (See exercise 9 this set.) Prove that $H \subset K$.

3. Let the finite group G have n distinct Sylow p-subgroups: $P_1, P_2, \ldots, P_n$. Prove that there exists a homomorphism $f : G \to S_n$, where S_n is the symmetric group on the n symbols: $P_1, P_2, P_3, \ldots, P_n$. Find the kernel of f in terms of the normalisers $N_G(P_i)$ of the P_i:

(i) in general,

(ii) when one of the Sylow p-subgroups is normal in G,

(iii) when $G = A_4$ and $p = 3$.

4. Let $I(G)$ be the group of inner automorphisms of G. Prove that $|I(G)| = 1$ if and only if G is abelian. Prove that any automorphism of S_3 is determined by its effect on the elements of order 2. Hence show that $A(S_3) = I(S_3)$, and find $A(S_3)$.

5. Let $G = \langle g \rangle$ be an **additive** cyclic group of order n. Prove

(i) if d is a positive integral divisor of n then the set of elements of G whose orders divide d forms a subgroup G_d of G of order d,

(ii) if t is a positive integer then tg generates G if and only if $(t, n) = 1$,

(iii) f is an automorphism of G if and only if $f(x) = tx$, for all $x \in G$, where t is a positive integer such that $(t, n) = 1$.

(Here (t, n) denotes the greatest common divisor of t and n.)

6. A group G of order 16 is defined by:

$$\langle a, b \mid a^4 = b^2 = c^2 = (ab)^2 = e,\ ac = ca,\ bc = cb \rangle .$$

Show that $G = AB$, where A and B are subgroups of order 4.

7. Let H be a subgroup of a group G. Let K be a finite cyclic subgroup of G. Let $H \subset K$ and $g^{-1}Hg \subset K$ for some $g \in G$. Prove that $H = g^{-1}Hg$.

8. Let K be a finite cyclic **normal** subgroup of a group G. Let H be any subgroup of K. Prove that H is normal in G.

9. Let G be the set of positive integers less than a given positive integer n and coprime with n. Show that G is an abelian group with respect to multiplication modulo n. Let $n = 16$. Find the orders of all the elements of G. Use the structure theory for abelian groups to identify G. Let m be any **odd** integer. Show that $m^4 - 1$ is divisible by 16.

(Integers x, y are **coprime** if and only if the G.C.D. $(x, y) = 1$. Multiplication modulo n means multiply in the ordinary way then give the remainder after division by n.)

10. Let G be a non-abelian group. Prove that $G/Z(G)$ cannot be cyclic. Hence show that $A(G)$ cannot be cyclic. (See exercise 4.12.)

11. Let G be a finite group. Prove that

(i) if G has less than 3 subgroups other than G and $\{e\}$ then G is cyclic,

(ii) if G has no subgroups other than G and $\{e\}$ then G has prime order,

(iii) if G has just one subgroup other than G and $\{e\}$ then $|G| = p^2$ for some prime p,

(iv) if G has exactly 2 subgroups other than G and $\{e\}$ then either $|G| = pq$ for distinct primes p and q, or $|G| = p^3$ for some prime p.

12. Let $H \subset K$ be subgroups of a finite group G. Prove that

$$[G : H] = [G : K][K : H] .$$

Let G be a finite group with two **distinct** subgroups H_1 and H_2 of index 3 in G. Prove that $H_1 \cap H_2$ has index either 6 or 9 in G.

Now let G be a group of order 51. Prove that G has a cyclic normal subgroup of order 17. Deduce that G is cyclic.

13. Let each element $x \neq e$ of a finite group G have order 2. Prove that G is isomorphic to a direct product of cyclic groups each of order 2. Hence determine all groups of order 4.

14. Let G be a finite group of order n. Let $g \in G$. Prove that $g^n = e$. By using the group G of integers $\{1, 2, 3, 4, \ldots, (p - 1)\}$ under multiplication modulo p, prove Fermat's theorem that $a^p \equiv a \pmod{p}$ for any integer a, where p is prime.

15. Let G be any group of order 24. Show that there are either 1 or 3 subgroups of order 8 and either 1 or 4 subgroups of order 3. Find the numbers of subgroups of orders 3 and 8 in

(i) C_{24} cyclic of order 24,

(ii) S_4,

(iii) $A_4 \times C_2$ (direct product).

For further exercises with answers the reader is referred to: 'Problems in Group Theory' — Dixon (Blaisdell, 1967) and 'Exercises in Group Theory' — Lyapin, Aizenshtat, and Lesokhin (Wolters—Noordhoff, Groningen, 1972).

OUTLINE SOLUTIONS TO THE EXERCISES

EXERCISES 1

1. Suppose e, e' both identifies then $e = ee' = e'$. Suppose b, b' both inverses of a, then

$$b = b(ab') = (ba)b' = b' .$$

2. Let $ba = e$. Then $(ab)(ab) = a(b(ab)) = a((ba)b) = a(eb) = ab$. Let x be left inverse of ab. Then $(x(ab))ab = x(ab)$. Thus

$$e(ab) = e .$$

Thus $$ab = e .$$

Let $a \in G$. Then $ea = a$. Choose b with $ba = e$. By above $ab = e$. Then

$$ae = a(ba) = (ab)a = ea = a .$$

Thus $$ae = a.$$

Hence G is a group according to section 1.2.

3. Let e be solution of $yb = b$ for some fixed b. Then $eb = b$. Let a be any element of G. Solve $bx = a$. Then

$$ea = ebx = bx = a .$$

Thus e is a left identity for G. The solution of $ya = e$ gives left inverse for a. Hence G is a group by 2 above.

4. (i) All invertible 2×2 matrices over $\mathbb{R}$ (see [G]).
(ii) $(\mathbb{Z}, +)$ integers under addition.
(iii) Symmetry group of equilateral triangle.
(iv) The set $\{1, -1, i, -i\}$ under ordinary multiplication of complex numbers. Here $i^2 = -1$.

EXERCISES 2

1. See Table 7.4.1, section 7.4.

2. Let $G = \{a_1, a_2, \ldots, a_n\}$ with $a_1 = e$. Let x appear twice in a row of the table. Then for some i, $j \neq k$, we have $a_i a_j = a_i a_k$. Then $a_i^{-1} a_i a_j = a_i^{-1} a_i a_k$. Thus $a_j = a_k$, $j \neq k$. This cannot be. Required result follows. Similarly for columns.

3. Because G is finite the sequence x, x^2, x^3, ... eventually repeats. Then $x^i = x^j$ for some i, j, $i < j$. Then $x^i . x^{-i} = x^j . x^{-i}$. Thus $e = x^{j-i}$. Put $r = j - i$, then $x^r = e$ with r a positive integer

4. Let $O(x) = n$, and $O(x^{-1}) = m$. Then $(x^{-1})^n = (x^n)^{-1} = e^{-1} = e$. Thus $m \leqq n$. Now $x = (x^{-1})^{-1}$; hence $n \leqq m$. Thus $m = n$.

5.

Element:	e	a	a^2	a^3	b	ab	a^2b	a^3b
Inverse:	e	a^3	a^2	a	b	ab	a^2b	a^3b
Order:	1	4	2	4	2	2	2	2

6. G cyclic $= \{e, a, a^2, \ldots, a^r\}$. Let $x, y \in G$. Then $x = a^i$, $y = a^j$ for some i, j. Thus $xy = a^{i+j} = yx$.

7. Klein 4-group $K_4 = \langle a, b \mid a^2 = b^2 = e, ab = ba \rangle$.

8. $f = \begin{pmatrix} 1 & 2 & \ldots & n \\ a_1 & a_2 & \ldots & a_n \end{pmatrix}$ Take first cycle to be:

$(1 \;\; f(1) \;\; f^2(1) \;\; f^3(1) \;\; \ldots \;\; f^i(1))$, where' $f^{i+1}(1) = 1$, $f^2(1) = f(f(1))$, etc. Take any m not in the first cycle. For second cycle take $(m \;\; f(m) \;\; f^2(m) \;\; \ldots \;\; f^j(m))$ where $f^{j+1}(m) = m$. Repeat. After finite number of steps process gives required result.

$$f = \begin{pmatrix} 1 & 2 & 3 & 4 & 5 & 6 & 7 \\ 3 & 4 & 1 & 6 & 7 & 2 & 5 \end{pmatrix} = (13)(246)(57)$$

$$g = \begin{pmatrix} 1 & 2 & 3 & 4 & 5 & 6 \\ 1 & 5 & 4 & 3 & 6 & 2 \end{pmatrix} = (1)(256)(34) .$$

9. $(a_1\ a_2\ a_3\ \ldots\ a_m) = (a_1\ a_m)(a_1\ a_{m-1})\ \ldots\ (a_1\ a_3)(a_1\ a_2)$.
Required result follows using question 8 above.

10. $O(f)$ = L.C.M. of 2, 3, 2 = 6

$O(g)$ = L.C.M. of 1, 3, 2 = 6 .

EXERCISES 3

1. Let $G = S_3 = \{(1), (12), (13), (23), (123), (132)\}$. Now $S = \{(12), (13)\}$ is **not** a subgroup of G. However, under the product $\times$ defined by : $(12) \times (13) = (13) \times (12) = (13)$, $(12) \times (12) = (12)$, $(13) \times (13) = (12)$, S is a cyclic group of order 2 with (12) acting as the identity.

2. Let H be a subgroup of the infinite cyclic group $C_\infty = \{e, a^{\pm 1}, a^{\pm 2}, a^{\pm 3}, \ldots\}$. Then H is cyclic $\{e, a^{\pm m}, a^{\pm 2m}, \ldots\}$.

Define a function $f : H \to C_\infty$ by: $f(a^{im}) = a^i$, $i \in \mathbf{Z}$. This is an isomorphism of H with C_∞.

3.

Figure 3.3.1

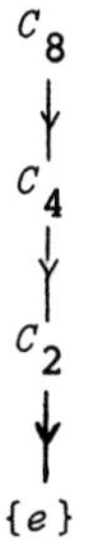

Figure 3.3.2

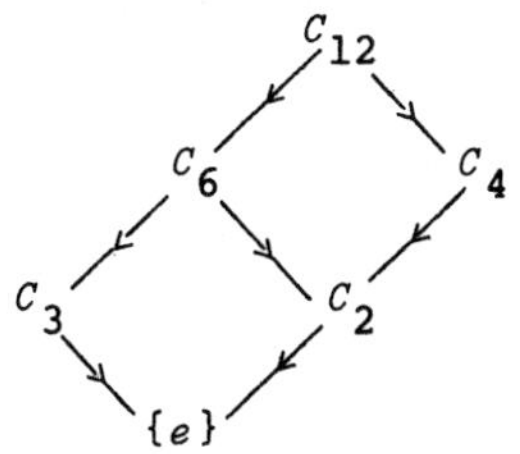

Figure 3.3.3

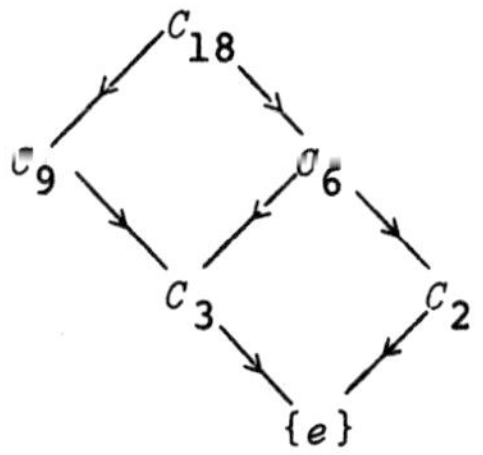

Figure 3.3.4

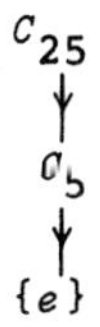

Figure 3.3.5

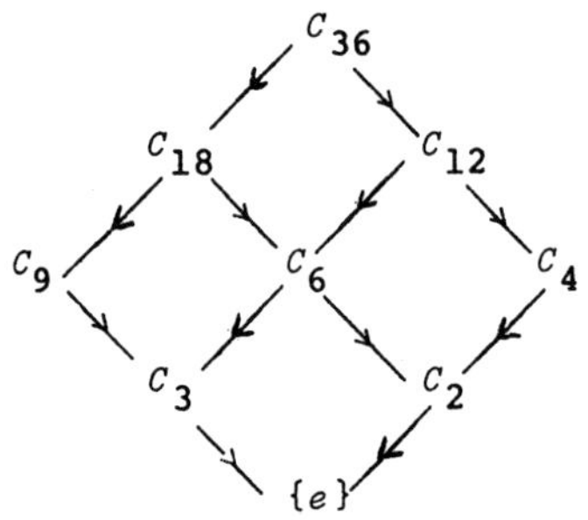

Figure 3.3.6

4. $(\underline{\mathbf{Z}}_p, +)$ is cyclic order p. By Lagrange's theorem the order of subgroup divides p. Thus subgroups have orders 1 or p. Thus subgroups are $(\underline{\mathbf{Z}}_p, +)$ itself and $\{\bar{0}\}$.

5. This is the multiplicative group $\underline{\mathbf{Z}}_7 - \{0\} = \underline{\mathbf{Z}}_7^*$ of the field $\underline{\mathbf{Z}}_7$.

	1	2	3	4	5	6
1	1	2	3	4	5	6
2	2	4	6	1	3	5
3	3	6	2	5	1	4
4	4	1	5	2	6	3
5	5	3	1	6	4	2
6	6	5	4	3	2	1

Table 3.5.1

G is associative because multiplication of integers is. The identity is 1. Inverses are as below:

Element:	1	2	3	4	5	6
Inverse:	1	4	5	2	3	6
Order:	1	3	6	3	6	2

Thus G is a group. Since $O(3) = 6$, $G = \langle 3 \rangle$ is cyclic order 6.

6. See Chapter 4 for right and left cosets of {(1), (12)} in S_3. For {(1), (13)} we have:

right cosets {(1), (13)}, {(12), (123)}, {(23), (132)}

left cosets {(1), (13)}, {(12), (132)}, {(23), (123)} .

For {(1), (23)} we have:

right cosets {(1), (23)}, {(12), (132)}, {(13), (123)}

left cosets {(1), (23)}, {(12), (123)}, {(13), (132)}

For {(1), (123), (132)} we have:

right and left cosets: {(1), (123), (132)}, {(12), (13), (23)} .

7. $$t\begin{pmatrix} 2 & 3 & 1 & 4 \\ 1 & 2 & 4 & 3 \end{pmatrix} = \frac{2-3}{1-2} \cdot \frac{2-1}{1-4} \cdot \frac{2-4}{1-3} \cdot \frac{3-1}{2-4} \cdot \frac{3-4}{2-3} \cdot \frac{1-4}{4-3}$$

$$= \left(\frac{-1}{-1}\right) \cdot \left(\frac{1}{-3}\right) \cdot \left(\frac{-2}{-2}\right) \cdot \left(\frac{2}{-2}\right) \cdot \left(\frac{-1}{-1}\right) \cdot \left(\frac{-3}{1}\right)$$

$$= -1 \quad .$$

8. $$t(g)t(f) = \Pi\left(\frac{b_i - b_j}{c_i - c_j}\right) \cdot \Pi\left(\frac{a_i - a_j}{b_i - b_j}\right)$$

$$= \Pi\left(\frac{a_i - a_j}{c_i - c_j}\right) = t(gf) \quad .$$

9. Transposition $f = (a_1 a_2)$ can be expressed

$$\begin{pmatrix} a_1 & a_2 & a_3 & a_4 & \cdots & a_n \\ a_2 & a_1 & a_3 & a_4 & \cdots & a_n \end{pmatrix} .$$

Thus $$t(f) = \frac{a_1 - a_2}{a_2 - a_1} \cdot \frac{a_1 - a_3}{a_2 - a_3} \cdot \frac{a_1 - a_4}{a_2 - a_4} \cdot \cdots \cdot$$

$$\frac{a_2 - a_3}{a_1 - a_3} \cdot \frac{a_2 - a_4}{a_1 - a_4} \cdot \cdots$$

$$= -1 \; .$$

10. $f = f_1 f_2 f_3 \dots f_m$ as a product of transpositions. Thus

$$t(f) = t(f_1 f_2 \dots f_m) = t(f_1)t(f_2) \dots t(f_m)$$

by question 8. Hence

$$t(f) = (-1)(-1) \dots (-1) = (-1)^m .$$

Thus $t(f) = \pm 1$ according as m is even or odd.

11. Let $x, y \in S$. Then xy is also even, hence $xy \in S$. By theorem 3.2.1 S is a subgroup of G.

12. By exercise 2.9 $t(a_1\ a_2\ a_3\ \dots\ a_r) = (-1)^{r-1}$. Thus

$$t(a_1 a_2) = -1$$

and $$t(a_1 a_2 a_3) = (-1)^2 = 1 .$$

Hence $$A_3 = \{(1), (123), (132)\} .$$

EXERCISES 4

1. As found in example 4.4.1 the conjugacy classes of S_3 are $\{(1)\}$, $\{(12), (13), (23)\}$, $\{(123), (132)\}$. By theorem 4.4.1 $\{(1)\}$, A_3 and S_3 are the only normal subgroups of S_3.

2. Let $g \notin H$. Then $G = H \cup gH = H \cup Hg$. Thus $gH = Hg$, if $g \notin H$. If $g \in H$ then $gH = H = Hg$.

3. (i) follows at once from theorem 4.3.1.

(ii) If $x \in H$ then $xH = H = Hx$; hence $x \in N_G(H)$. Thus $H \subset N_G(H)$. Also if $x \in N_G(H)$ then $xH = Hx$. Thus H is normal in $N_G(H)$. This gives (ii).

(iii) If H is normal in N, then $x \in N \rightarrow xH = Hx \rightarrow x \in N_G(H)$. Thus $N \subset N_G(H)$, which is (iii).

4. (i) $xx^{-1} = e \in H \rightarrow xRx$.

(ii) $xy^{-1} \in H \rightarrow (xy^{-1})^{-1} \in H \rightarrow yx^{-1} \in H$. Thus $xRy \rightarrow yRx$.

(iii) $xy^{-1} \in H$, $yz^{-1} \in H \rightarrow xy^{-1}yz^{-1} \in H \rightarrow xz^{-1} \in H$. Thus xRy, $yRz \rightarrow xRz$. Hence R is an equivalence relation. Let $H(x)$ be the equivalence class containing x. Let $y \in H(x)$. Then $yRx \rightarrow yx^{-1} \in H \rightarrow y \in Hx \rightarrow H(x) \subset Hx$. Conversely $y \in Hx \rightarrow y = hx$, $h \in H \rightarrow yx^{-1} = h \rightarrow yx^{-1} \in H \rightarrow yRx \rightarrow y \in H(x) \rightarrow Hx \subset H(x)$. Altogether $H(x) = Hx$.

5. Similar to 4 above.

6. Let $f = c_1 c_2 \ldots c_r$ as a product of disjoint cycles. Since $h^{-1}fh = (h^{-1}c_1h) \ldots (h^{-1}c_rh)$, it suffices to look at one cycle. Suppose that $c_i = (a_1\ a_2 \ldots a_s)$ is a cycle of length s. Let

$$h^{-1} = \begin{pmatrix} 1 & 2 & 3 & \ldots & n \\ 1' & 2' & 3' & \ldots & n' \end{pmatrix}$$

and

$$h = \begin{pmatrix} 1' & 2' & 3' & \ldots & n' \\ 1 & 2 & 3 & \ldots & n \end{pmatrix}.$$

Then $h^{-1}c_ih = (a_1'\ a_2' \ldots a_s')$. Thus $h^{-1}fh$ has the same cycle pattern as f. Conversely suppose f and $g \in S_n$ have the same cycle pattern. We have:

$$f = (a_{11}\ a_{12} \ldots a_{1s_1})(a_{21}\ a_{22} \ldots a_{2s_2}) \ldots (a_{t1}\ a_{t2} \ldots a_{ts_t})$$

$$g = (b_{11}\ b_{12} \ldots b_{1s_1})(b_{21}\ b_{22} \ldots b_{2s_2}) \ldots (b_{t1}\ b_{t2} \ldots b_{ts_t})$$

Put

$$h = \begin{pmatrix} b_{11} & b_{12} & \ldots & b_{1s_1} & b_{21} & b_{22} & \ldots & b_{2s_2} & \ldots & b_{t1} & b_{t2} & \ldots & b_{ts_t} \\ a_{11} & a_{12} & \ldots & a_{1s_1} & a_{21} & a_{22} & \ldots & a_{2s_2} & \ldots & a_{t1} & a_{t2} & \ldots & a_{ts_t} \end{pmatrix}$$

Then $h^{-1}fh = g$.

7. Let f have cycle pattern: $(n_1, n_2, \ldots, n_r)$. The permutations conjugate to f are precisely those with the same cycle pattern. There are $n!$ ways of writing such a pattern in the order: (cycles length 1)(cycles length 2) ...; **but with repetitions** which occur as follows. Each set of n_i cycles of length i can be written in any order in $n_i!$ ways. Each cycle of length i can be written in i different ways, e.g. $(a_1\ a_2\ a_3 \ldots a_i) = (a_i\ a_1\ a_2 \ldots a_{i-1})$. Thus each permutation conjugate to f has been written $n_i!\,i^{n_i}$ times. Thus the number required is:

$$n! \Big/ (1^{n_1} n_1!\ 2^{n_2} n_2! \ldots r^{n_r} n_r!) .$$

8. For (12)(3) cycle pattern is: (1, 1). Number of conjugates to (12)(3) is: $3!/(1^1\ 1!\ \ 2^1\ 1!) = \frac{6}{2} = 3$. Similarly for the others.

9. Composition of functions is associative. The identity function $e(g) = g$, for all $g \in G$, is an automorphism of G. Thus $A(G)$ has an identity element. If $f \in A(G)$ then f^{-1} exists as a **function**. Consider $f^{-1}(xy)$; $x, y \in G$. Let $x = f(a)$, $y = f(b)$ for some $a, b \in G$. Then $f^{-1}(x) = a$, $f^{-1}(y) = b$, and

$$f^{-1}(xy) = f^{-1}(f(a)f(b)) = f^{-1}(f(ab)) = ab = f^{-1}(x)f^{-1}(y).$$

Thus $f^{-1} \in A(G)$. By section 1.2 $A(G)$ is a group.

10. $i_x(g_1) = i_x(g_2) \to xg_1x^{-1} = xg_2x^{-1} \to g_1 = g_2 \to i_x$ one-one.

If $g \in G$ then $i_x(x^{-1}gx) = g \to i_x$ onto.

$$i_x(ab) = xabx^{-1} = xax^{-1}xbx^{-1} = i_x(a)i_x(b).$$

Thus $i_x \in A(G)$.

11. Let $i_x, i_y \in I(G)$. Then $i_x i_y^{-1} = i_{xy^{-1}} \in I(G)$. By theorem 3.2.1 $I(G)$ is a subgroup of $A(G)$. Let $f \in A(G)$ and $i_x \in I(G)$. Then $f^{-1}i_x f = i_{f^{-1}(x)} \in I(G)$. Thus $I(G)$ is normal in $A(G)$ by theorem 4.1.2.

12. Define function $f\colon G \to I(G)$ by $f(x) = i_x$; $x \in G$. Then $f(xy) = i_{xy} = i_x i_y = f(x)f(y)$. Also f is onto. By theorem 4.5.3 $I(G) \cong G/\mathrm{Ker}\, f$. Now $x \in \mathrm{Ker}\, f \leftrightarrow f(x) = i_x$ = identity on $G \leftrightarrow xgx^{-1} = g$ for all $g \in G \leftrightarrow x \in Z(G)$. Thus $\mathrm{Ker}\, f = Z(G)$. Hence

$$I(G) \cong G/Z(G).$$

13. $i_x(H) \subset H \leftrightarrow xHx^{-1} \subset H \leftrightarrow H$ is normal by theorem 4.1.2, where x is any element of G.

14. (i) Let $O(g) = m$. Let $O(x^{-1}gx) = n$. Then

$$(x^{-1}gx)^m = x^{-1}gx.x^{-1}gx \ldots x^{-1}gx = x^{-1}g^m x = x^{-1}ex = e .$$

Thus $n = O(x^{-1}gx) \leq m$. Also

$$e = (x^{-1}gx)^n = x^{-1}g^n x \to g^n = e \to O(g) \leq n .$$

Thus $n \leq m$ and $m \leq n$. Hence $m = n$.

(ii) Let H be normal in G. Let x be conjugate to an element $h \in H$. Then $x = g^{-1}hg \in H$. Thus H consists of complete conjugacy classes. Conversely suppose H consists of complete conjugacy classes. Let $h \in H$, $g \in G$. Then $g^{-1}hg$ is conjugate to h hence lies in the same class as h. Now this whole class is a subset of H. Thus $g^{-1}hg \in H$. By theorem 4.1.2 H is normal in G.

15. If G has no odd permutations then $G = S$. If G has some odd permutations define a function $f\colon G \to \{1, -1\}$, where the latter is a cyclic group of order 2 under ordinary multiplication of integers, by $f(g) = 1$ if g is even,

$$= -1 \text{ if } g \text{ is odd.}$$

Because G has odd permutations f is onto. By 4.5.3 $C_2 \cong G/\mathrm{Ker}\, f$. Now $\mathrm{Ker}\, f = S$. Thus $G/S \cong C_2$. Hence $|G| = |S||C_2|$. Thus $|S| = |G|/2$.

16. $|A_4| = 12$. Any subgroup H order 6 is of index 2 and hence normal. Thus H must be the union of complete conjugacy classes by theorem 4.4.1. Calculation shows that A_4 has 4 conjugacy classes with 1, 3, 4, 4 elements respectively. There is no way of adding these numbers to give 6. Hence no subgroup H of order 6 can exist.

17. By question 16 above, 6 divides $|A_4|$ but A_4 has no subgroup of order 6.

18. Cycle patterns for S_4 are (4), (2, 1), (1, 0, 1), (0, 2), (0, 0, 0, 1). Corresponding conjugacy classes are:

{1}, {(12), (13), (14), (23), (24), (34)},

{(123), (124), (132), (134), (143), (142), (234), (243)},

{(12)(34), (13)(24), (14)(23)}, {(1234), (1324), (1342), (1243), (1432), (1423)} .

A normal subgroup must be a union of {(1)} with some of the other classes to form a subgroup whose order divides 24. Numbers in the classes are: 1, 6, 8, 3, 6. Possibilities are: 1, 1 + 3, 1 + 8 + 3, 24. These correspond to

{(1)}, {(1), (12)(34), (13)(24), (14)(23)} $\cong K_4$, A_4 and S_4.

Thus the normal subgroups of S_4 are $\{(1)\}$, K_4, A_4, and S_4, of orders 1, 4, 12, and 24.

19. A normal subgroup of A_5 must be a union of $\{(1)\}$ with other conjugacy classes of A_5 to form a subgroup whose order divides $|A_5| = 60$. Add 1 to 20, 15, 12, in various combinations and see if the sum divides 60. Only possibilities are 1 and 60 corresponding to $\{(1)\}$ and A_5. Hence A_5 is simple.

20. G simple abelian. Let $e \neq g \in G$. Then $\langle g \rangle$ is a cyclic subgroup of G. Moreover $\langle g \rangle$ is normal because G is abelian. Hence $\langle g \rangle = G$. Now G has only the subgroups $\{e\}$ and G. By theorem 3.6.1 G has prime order.

21. Let S be the set of all even permutations in G. By question 15 above S is normal in G and $S \neq G$ because G contains an odd permutation. If $S = \{e\}$ then $|G|/2 = |S| = 1$ and $G \cong C_2$, otherwise $S \neq \{e\}$ and then G is not simple.

22. Let $x \in G(s) \cap G(t)$. Then $x = f(s) = h(t)$; $f, h \in G$. Let $y \in G(s)$ then $y = g(s)$ for some $g \in G$. Thus $y = gf^{-1}h(t) \in G(t)$. Hence $G(s) \subset G(t)$. Similarly $G(t) \subset G(s)$. Thus $G(t) = G(s)$.

23. Let $f, g \in G_s$. Then $fg^{-1}(s) = s$. Thus $fg^{-1} \in G_s$. By theorem 3.2.1 G_s is a subgroup of G.

24. Let $g \in G_s$. Then $fgf^{-1}(t) = fg(s) = f(s) = t$. Thus $fgf^{-1} \in G_t$. Hence $fG_sf^{-1} \subset G_t$. Let $h \in G_t$. Then $f^{-1}hf(s) = f^{-1}h(t) = f^{-1}(t) = s$. Thus $f^{-1}hf \in G_s$. Hence $f^{-1}G_tf \subset G_s$. Thus $G_t \subset fG_sf^{-1}$. Altogether $fG_sf^{-1} = G_t$, where $f(s) = t$.

25. Let $G = G_s \cup g_1G_s \cup \ldots \cup g_rG_s$ be the left coset decomposition of G relative to G_s. $r = |G|/|G_s|$. Let $x \in g_iG_s$. Then $x(s) = g_i(s)$. Thus every element of the coset g_iG_s sends s into the same element of $S = \{1, 2, \ldots, n\}$, namely $g_i(s)$. On the other hand $x(s) = y(s) \to y^{-1}x(s) = s \to y^{-1}x \in G_s \to x \in yG_s \to x$ and y are in the same left coset yG_s. Hence $|G(s)| = r$, the number of cosets. Thus

$$|G(s)| = |G|/|G_s|.$$

Hence

$$|G| = |G_s||G(s)|.$$

26. $S_3/A_3 = \{\bar{e}, \overline{(12)}\} \cong C_2$, where $\bar{e} = A_3$, $\overline{(12)} = (12)A_3 = \{(12), (13), (23)\}$. Subgroups of S_3/A_3 are $A_3/A_3 = \{\bar{e}\}$ and S_3/A_3.

27. $S_4/K_4 = S_4/\{(1), (12)(34), (13)(24), (14)(23)\}$ by question 18.

Thus $S_4/K_4 = \{\overline{(1)}, \overline{(12)}, \overline{(13)}, \overline{(23)}, \overline{(123)}, \overline{(132)}\}$,

where
$$\begin{aligned}
\bar{e} = \overline{(1)} &= \{(1), (12)(34), (13)(24), (14)(23)\} = K_4 \\
\overline{(12)} &= \{(12), (34), (1324), (1423)\} = (12)K_4 \\
\overline{(13)} &= \{(13), (1234), (24), (1432)\} = (13)K_4 \\
\overline{(23)} &= \{(23), (1243), (1342), (14)\} = (23)K_4 \\
\overline{(123)} &= \{(123), (134), (243), (142)\} = (123)K_4 \\
\overline{(132)} &= \{(132), (234), (124), (143)\} = (132)K_4
\end{aligned}$$

Put $a = \overline{(123)}$, $b = \overline{(12)}$. Then S_4/K_4 is a group of order 6 generated by a and b subject to $a^3 = b^2 = e$, $ba = a^2b$. Sections 2.5 and 2.6 show that $S_4/K_4 \cong S_3$. In fact we can identify S_4/K_4 with S_3 as follows:

$$\begin{array}{rcl}
S_4/K_4 &=& \{\overline{(1)}, \overline{(12)}, \overline{(13)}, \overline{(23)}, \overline{(123)}, \overline{(132)}\} \\
&& \quad\updownarrow \quad\;\; \updownarrow \quad\;\; \updownarrow \quad\;\; \updownarrow \quad\;\;\; \updownarrow \quad\;\;\; \updownarrow \\
S_3 &=& \{(1), (12), (13), (23), (123), (132)\}
\end{array}$$

<u>Note</u> that the elements of S_4 mapping onto $\bar{a}$ in the factor group S_4/K_4 under the natural epimorphism $S_4 \to S_4/K_4$ are precisely the elements of the coset aK_4. This gives the correspondence between subgroups of S_4/K_4 and subgroups of S_4 containing K_4 as follows:

$$\begin{array}{cccccc}
\{(1)\}, & \{(1), (12)\}, & \{(1), (13)\}, & \{(1), (23)\}, & \{(1), (123), (132)\}, & S_4/K_4 \\
\updownarrow & \updownarrow & \updownarrow & \updownarrow & \updownarrow & \updownarrow \\
K_4, & K_4 \cup (12)K_4, & K_4 \cup (13)K_4, & K_4 \cup (23)K_4, & K_4 \cup (123)K_4 \cup (132)K_4, & S_4
\end{array}$$

28. Using the results in question 27 above we have:

$A_4/K_4 = \{\overline{(1)}, \overline{(123)}, \overline{(132)}\} \cong A_3 = \{(1), (123), (132)\} \cong C_3$.

The correspondence between subgroups of A_4/K_4 and subgroups of A_4 containing K_4 is given by:

$$\begin{array}{ccc}
\{(1)\} & \qquad & A_4/K_4 \\
\updownarrow & & \updownarrow \\
K_4 & & A_4 = K_4 \cup (123)K_4 \cup (132)K_4
\end{array}$$

29. $f\colon G \to M$. Let $x \in f^{-1}(f(H))$. Then $f(x) \in f(H) \to f(x) = f(h)$ for some $h \in H \to f(xh^{-1}) = e \to xh^{-1} \in \text{Ker} f \to x \in (\text{Ker} f)H \to f^{-1}f(H)) \subset (\text{Ker} f)H$. Conversely let $x \in (\text{Ker} f)H$. Then we have:

$x = yh$; $y \in \text{Ker} f$, $h \in H \to f(x) = f(y)f(h) = f(h) \in f(H) \to x \in f^{-1}(f(H)$

Thus $\quad (\text{Ker} f)H \subset f^{-1}(f(H))$.

Altogether

$$f^{-1}(f(H)) = (\text{Ker} f)H = H(\text{Ker} f).$$

30. $f\colon G \to M$. Let $x, y \in f(H)$. Then $x = f(g)$, $y = f(h)$, for some $g, h \in H$. Then $xy^{-1} = f(gh^{-1}) \in f(H)$. By theorem 3.2.1 $f(H)$ is a subgroup of M. Let H be normal in G. Let $x \in f(H)$, $y \in f(G)$. Then $x = f(h)$, $y = f(g)$ for some $h \in H$, $g \in G$. Thus $y^{-1}xy = f(g^{-1}hg) \in f(H)$ because $g^{-1}hg \in H$ (normal). Hence $f(H)$ is normal in $f(G)$ by theorem 4.1.2. Let L be subgroup of M. Let $g, h \in f^{-1}(L)$. Then $f(g), f(h) \in L$. Thus $f(gh^{-1}) = f(g)[f(h)]^{-1} \in L$. Thus $gh^{-1} \in f^{-1}(L)$. By theorem 3.2.1 $f^{-1}(L)$ is a subgroup of G. Let L be normal in M. Let $g \in G$, $h \in f^{-1}(L)$. Then

$$f(g^{-1}hg) = [f(g)]^{-1}f(h)f(g) \in f(g)^{-1}Lf(g) \subset L$$

since L is normal in M. Thus $g^{-1}hg \in f^{-1}(L)$. By theorem 4.1.2 $f^{-1}(L)$ is normal in G.

31. $f\colon G \to M$. Let f be one-one. Let $g \in \text{Ker} f$. Then $f(g) = e = f(e)$. Now f is one-one; hence $g = e$. Conversely let $\text{Ker} f = \{e\}$. Let $f(g) = f(h)$. Then

$$f(gh^{-1}) = e\ .$$

Thus $gh^{-1} \in \text{Ker} f = \{e\}$. Hence

$$gh^{-1} = e\ .$$

Thus $g = h$. Hence f is one-one.

32. $f(xy) = i_{xy} = i_x i_y = f(x)f(y)$.

The orbits are just the conjugacy classes of G. By theorem 4.6.4, the number of conjugacy classes

$$t = \frac{1}{|G|} \sum_{x \in G} |F(x)|\ .$$

Now

$$F(x) = \{g \in G \mid f(x)(g)\} = g\} = \{g \in G \mid xgx^{-1} = g\} = N_G(x).$$

Thus

$$t = \frac{1}{|G|} \sum_{x \in G} |N_G(x)| = \sum_{x \in G} \frac{|N_G(x)|}{|G|}$$

$$= \sum_{x \in G} \frac{1}{[G : N_G(x)]}.$$

EXERCISES 5

1. $G = A \oplus B$. Define a function f: $G \to B$ by $f(a + b) = b$, where $a \in A$, $b \in B$. This is a function because each $g \in G$ has a **unique** expression in the form $g = a + b$. f is an epimorphism. By theorem 4.5.3 $G/\text{Ker}\, f \cong B$. Now $\text{Ker}\, f = A$. Hence $G/A \cong B$. Similarly $G/B \cong A$.

2. Associativity follows from that in each G_i. Identity is $(e_1, e_2, \ldots, e_n)$ where e_i is the identity of G_i. Inverse of $(g_1, g_2, \ldots, g_n)$ is $(g_1^{-1}, g_2^{-1}, \ldots, g_n^{-1})$. Thus G is a group.

3. Let f: $G_1 \times G_2 \times \ldots \times G_n \to G$ be an isomorphism. Identify G_i with the subgroup $\{(e_1, e_2, \ldots, e_{i-1}, g_i, e_{i+1}, \ldots, e_n) \mid g_i \in G_i\}$ of G. Put $H_i = f(G_i)$. Then (a), (b), (c), and (d) follow for H_i from the corresponding properties of the G_i as subgroups of G. Conversely let G contain subgroups H_i, ($i = 1, 2, \ldots, n$), with the given properties (a), (b), (c), (d). We identify H_i and G_i and define a function f: $G_1 \times G_2 \times \ldots \times G_n \to G$ by $f(g_1, g_2, \ldots, g_n) = g_1 g_2 \ldots g_n$. By condition (c) f is onto. Consider $g_i \in G_i$ and $g_j \in G_j$, $i \neq j$. $x = g_i^{-1}(g_j^{-1} g_i g_j) \in G_i$, and $x = (g_i^{-1} g_j^{-1} g_i) g_j \in G_j$, by (b). Thus

$$x = [g_i, g_j] \in G_i \cap G_j = \{e\}$$

by (d). Thus

$$g_i g_j = g_j g_i .$$

Hence elements from different subgroups G_i commute. Suppose

$$f(g_1, g_2, \ldots, g_n) = f(h_1, h_2, \ldots, h_n); \quad h_i, g_i \in G_i.$$

Then $\quad g_1 g_2 \ldots g_n = h_1 h_2 \ldots h_n$. Thus

$$h_1^{-1} g_1 = (h_2 h_3 \ldots h_n)(g_2 g_3 \ldots g_n)^{-1} = h_2 h_3 \ldots h_n . g_n^{-1} \ldots g_2^{-1}$$

$$= h_2 g_2^{-1} . h_3 g_3^{-1} \ldots h_n g_n^{-1}$$

by the result that $g_i g_j = g_j g_i$, $i \neq j$. By (d) $h_1^{-1} g_1 = e$. Thus

$$g_1 = h_1 .$$

Repeat for $g_2 g_3 \ldots g_n = h_2 h_3 \ldots h_n$. Eventually $h_i = g_i$, $(i = 1, 2, \ldots, n)$. Thus f is one-one.

$$f((g_1, g_2, \ldots, g_n)(h_1, h_2, \ldots, h_n)) = f(g_1 h_1, g_2 h_2, \ldots, g_n h_n)$$

$$= g_1 h_1 g_2 h_2 \ldots g_n h_n = (g_1 g_2 \ldots g_n)(h_1 h_2 \ldots h_n)$$

$$= f((g_1, g_2, \ldots, g_n)) f((h_1, h_2, \ldots, h_n))$$

using $g_i h_j = h_j g_i$, $i \neq j$. Thus f is a homomorphism. Altogether f is an isomorphism.

4. Assume (a), (b), (c), (d) of question 3. We have proved above that $h_i h_j = h_j h_i$ for any $h_i \in H_i$ and $h_j \in H_j$. From (c) we can write $g \in G$ in form: $g = h_1 h_2 \ldots h_n$. Suppose this can be done in two ways:

$$h_1 h_2 \ldots h_n = g_1 g_2 \ldots g_n; \quad h_i, g_i \in H_i, \ (i = 1, 2, \ldots, n).$$

Then, as in question 3, $h_i = g_i$; $(i = 1, 2, \ldots, n)$.
Conversely assume (a) $h_i h_j = h_j h_i$; $h_i \in H_i$, $h_j \in H_j$, $i \neq j$.

(b) $g \in G$ can be written **uniquely** apart from order in the form:

$$g = h_1 h_2 \ldots h_n , \quad h_i \in H_i.$$

Let $h_i \in H_i$, $g \in G$. Write $g = g_1 g_2 \ldots g_n$, $g_j \in H_j$.

Then $\quad g^{-1} h_i g = g_n^{-1} g_{n-1}^{-1} \ldots g_i^{-1} g_{i-1}^{-1} \ldots g_1^{-1} h_i g_1 g_2 \ldots g_n$

$= g_n^{-1} g_{n-1}^{-1} \ldots (g_i^{-1} h_i g_i) \ldots g_n$, using (a),

$= (g_i^{-1} h_i g_i) . g_n^{-1} \ldots g_{i+1}^{-1} g_{i+1} \ldots g_n$, using (a),

$= g_i^{-1} h_i g_i \in H_i$.

By theorem 4.1.2 H_i is normal in G. This proves question 3 (b). Question 3 (c) follows at once from (b). Let $x \in H_i \cap H_1 H_2 \ldots H_{i-1} H_{i+1} \ldots H_n$. Then

$$x = h_i = h_1 h_2 \ldots h_{i-1} e\, h_{i+1} \ldots h_n$$

for suitable $h_i \in H_i$; $(i = 1, 2, 3, \ldots, n)$. Then

$$eee \ldots eh_i e \ldots e = h_1 h_2 \ldots h_{i-1} e\, h_{i+1} \ldots h_n.$$

By (b) $h_i = e$ for $i = 1, 2, \ldots, n$. In particular $x = e$. This gives question 3 (d).

5. $h_1 = u_1 + u_2 + 10u_3$, $h_2 = u_1 + 10u_2 + u_3$, $h_3 = 10u_1 + u_2 + u_3$ are linearly independent over $\mathbf{Z}$ and

$$A \begin{pmatrix} u_1 \\ u_2 \\ u_3 \end{pmatrix} = \begin{pmatrix} h_1 \\ h_2 \\ h_3 \end{pmatrix}$$

where $A = \begin{pmatrix} 1 & 1 & 10 \\ 1 & 10 & 1 \\ 10 & 1 & 1 \end{pmatrix}$. Use elementary row and column operations as given in the text to change A to the form:

$$\begin{pmatrix} \gamma_1 & 0 & 0 \\ 0 & \gamma_2 & 0 \\ 0 & 0 & \gamma_3 \end{pmatrix}$$

with $\gamma_1 \mid \gamma_2$ and $\gamma_2 \mid \gamma_3$.

We have the following sequence:

$$\begin{pmatrix} 1 & 1 & 10 \\ 1 & 10 & 1 \\ 10 & 1 & 1 \end{pmatrix} \to \begin{pmatrix} 1 & 0 & 0 \\ 1 & 9 & -9 \\ 10 & -9 & -99 \end{pmatrix} \to \begin{pmatrix} 1 & 0 & 0 \\ 0 & 9 & -9 \\ 0 & -9 & -99 \end{pmatrix} \to \begin{pmatrix} 1 & 0 & 0 \\ 0 & 9 & 0 \\ 0 & -9 & -108 \end{pmatrix} \to$$

$$\to \begin{pmatrix} 1 & 0 & 0 \\ 0 & 9 & 0 \\ 0 & 0 & 108 \end{pmatrix}.$$

Thus $F/H \cong C_1 \oplus C_9 \oplus C_{108}$; $C_1 = \{0\}$ can be discarded. Now

$108 = 3^3 \times 2^2$. Thus $C_{108} \cong C_{27} \oplus C_4$. Finally $F/H \cong C_9 \oplus C_{27} \oplus C_4$ in terms of indecomposable cyclic groups.

$$|F/H| = 9 \times 27 \times 4 = 972.$$

6. $G \cong F/H$ where F is free on a, b, c, d and H is generated by $h_1 = 11a - 10b - 4c - 7d$, $h_2 = 3a - 4b + 2c + 5d$, $h_3 = 3a - 4b - 4c - 7d$, which are linearly independent over $\mathbb{Z}$ and therefore form a basis of H.

$$A\begin{pmatrix} a \\ b \\ c \\ d \end{pmatrix} = \begin{pmatrix} h_1 \\ h_2 \\ h_3 \end{pmatrix},$$

where $A = \begin{pmatrix} 11 & -10 & -4 & -7 \\ 3 & -4 & 2 & 5 \\ 3 & -4 & -4 & -7 \end{pmatrix}$. A reduction as for question 5 leads to the matrix:

$$\begin{pmatrix} \gamma_1 & 0 & 0 & 0 \\ 0 & \gamma_2 & 0 & 0 \\ 0 & 0 & \gamma_3 & 0 \end{pmatrix} = \begin{pmatrix} 1 & 0 & 0 & 0 \\ 0 & 2 & 0 & 0 \\ 0 & 0 & 6 & 0 \end{pmatrix}.$$

<u>Note</u> that if h_1, h_2, h_3 had been linearly dependent over $\mathbb{Z}$ then one of the γ_i would have turned out to be 0. Thus

$$G \cong C_1 \oplus C_2 \oplus C_6 \oplus C_\infty .$$

Since $6 = 2 \times 3$, $C_6 \cong C_2 \oplus C_3$. Also $C_1 = \{0\}$ and can be discarded.

Finally $G \cong C_2 \oplus C_2 \oplus C_3 \oplus C_\infty$.

EXERCISES 6

1. $|G| = 20 = 2^2 \times 5$. $(1 + 5k)$ Sylow 5-subgroups each of order 5. Also $(1 + 5k) \mid 20$. Thus only one Sylow 5-subgroup. Normal by theorem 6.3.4.

2. $|G| = 36 = 2^2 \times 3^2$. $(1 + 2k)$ Sylow 2-subgroups each order 4 and $(1 + 2k) \mid 36$. $(1 + 3k)$ Sylow 3-subgroups each order 9 and $(1 + 3k) \mid 36$. Thus 1 or 4 subgroups of order 9. If only 1 sub-

group of order 9 then it is normal and G is not simple. If 4 subgroups of order 9. Let H, K be any 2 of them. Since $|H \cap K|$ divides $|H| = 9$, $|H \cap K| = 1, 3, 9$. If $|H \cap K| = 1$, then by theorem 3.5.2(3)

$$|HK| = |H||K|/|H \cap K| = 81 > |G| .$$

Contradiction. Since H, K are distinct, $|H \cap K| \neq 9$. Thus $|H \cap K| = 3$. Let $N = N_G(H \cap K)$. By theorem 4.9.1 H and K are abelian. Thus $H \subset N$ and $K \subset N$. Hence $HK \subset N$ and $|HK| = |H||K|/|H \cap K| = 81/3 = 27$. Thus $27 \leq |N| \leq 36$ and $|N|$ divides $|G| = 36$. Hence $|N| = 36$. Thus $N = G$. Hence $H \cap K$ is normal in G. Altogether G has at least one normal subgroup when $|G| = 36$.

3. $|G| = 48 = 2^4 \times 3$. $(1 + 3k)$ subgroups order 3. $(1 + 3k) \mid 48$. $(1 + 2k)$ subgroups order 16. $(1 + 2k) \mid 48$. Thus 1 or 3 Sylow 2-subgroups each of order 16. If 1 then it is normal by theorem 6.3.4 and G is not simple. If 3, let H, K be any 2 of them. Since $|H \cap K|$ divides $|H| = 16$, $|H \cap K| = 1, 2, 4, 8, 16$. Because H and K are distinct $|H \cap K| \neq 16$. If $|H \cap K| \leq 4$,

$$|HK| = |H||K|/|H \cap K| \geq \frac{16 \times 16}{4} = 64 > |G|.$$

Contradiction. Thus $|H \cap K| = 8$. Thus $H \cap K$ has index 2 in both H and K and so is normal in both. Let $N = N_G(H \cap K)$. Thus $H \subset N$ and $K \subset N$; hence $HK \subset N$. Now

$$|HK| = |H||K|/|H \cap K| = 32.$$

Thus $32 \leq |N| \leq 48$, and $|N|$ divides $|G| = 48$. Thus $|N| = 48$. Hence $N = G$. Hence $H \cap K$ is normal in G. Altogether any G of order 48 is not simple.

4. $|G| = 200 = 2^2 \times 5^2$. $(1 + 2k)$ subgroups of order 4. $(1 + 2k) \mid 200$. $(1 + 5k)$ subgroups of order 25. $(1 + 5k) \mid 200$. Thus only 1 subgroup of order 25 which must be normal. Hence G is not simple.

5. $|G| = 195 = 3 \times 5 \times 17$. $(1 + 3k)$ subgroups of order 3 $(1 + 3k) \mid 255$. $(1 + 5k)$ subgroups of order 5. $(1 + 5k) \mid 255$. $(1 + 17k)$ subgroups of order 17. $(1 + 17k) \mid 255$. Hence G has just

one normal subgroup H of order 17. $|G/H| = 3 \times 5$. Thus G/H is cyclic by theorem 6.4.1. By theorem 4.5.4 $G' \subset H$. Hence $|G'| = 1$ or 17. G has 1 or 85 subgroups of order 3 and 1 or 51 subgroups of order 5. If G has 85 subgroups of order 3 and 51 of order 5, since these can intersect only in $\{e\}$, G must have at least $(2 \times 85) + (4 \times 51) + 1 = 375$ elements. Contradiction. Thus G has a normal subgroup K of order 3 or 5. Then $|G/K| = 3 \times 17$ or 5×17. In either case G/K is cyclic by theorem 6.4.1. By theorem 4.5.4 $G' \subset K$. Thus $|G'| = 1$, 3, or 5. However, by the above, $|G'| = 1$ or 17. Thus $|G'| = 1$. Thus G is abelian. By the structure theory Chapter 5

$$G \cong C_3 \oplus C_5 \oplus C_{17}.$$

Let $C_3 = \langle a \rangle$, $C_5 = \langle b \rangle$, $C_{17} = \langle c \rangle$. Then by applying theorem 2.8.1 (5) twice we get

$$O(abc) = 3 \times 5 \times 17 = 255 .$$

Thus $G = \langle (abc) \rangle$ is cyclic.

6. $|A_4| = 12 = 2^2 \times 3$. Sylow's third theorem gives 1 or 3 subgroups order 4 and 1 or 4 subgroups order 3. $K_4 = \{(1), (12)(34), (13)(24), (14)(23)\}$ is **normal** in A_4, see exercise 4.18. Thus this must be the only subgroup of order 4. Subgroups order 3 are cyclic, generated by elements of order 3. By exercise 4.18 there are 4 subgroups of order 3:

$\{(1), (123), (132)\}$, $\{(1), (124), (142)\}$,

$\{(1), (134), (143)\}$, $\{(1), (234), (243)\}$.

7. $|G| = n = p_1^{t_1} p_2^{t_2} \dots p_r^{t_r}$, with $p_1, p_2, \dots, p_r$ distinct primes. By theorem 6.3.4 G has precisely one normal subgroup G_i of order $p_i^{t_i}$; $(i = 1, 2, \dots, n)$, and moreover $G_i \cap G_j = \{e\}$ by Lagrange's theorem. Let $g \in G_i$ and $h \in G_j$, $i \neq j$. Then

$$[h, k] = h^{-1}k^{-1}hk \in G_i \cap G_j$$

by normality of G_i and G_j. Thus $[h, k] = e$ which means $hk = kh$. Hence elements from **different** Sylow subgroups commute. Let $x \in G_i \cap (G_1 G_2 \dots G_{i-1} G_{i+1} \dots G_r)$. Then $x \in G_i$. Hence order

$x = O(x) = p_i^{s_i}$, $0 \leqq s_i \leqq t_i$. Also $x \in G_1 G_2 \ldots G_{i-1} G_{i+1} \ldots G_r$.
Thus

$$x = g_1 g_2 \ldots g_{i-1} g_{i+1} \ldots g_r$$

with $g_j \in G_j$, $(j = 1, 2, \ldots, i-1, i+1, \ldots, r)$. Put $\ell = n/p_i^{t_i}$, so that $(\ell, p_i) = 1$. Then

$$x^\ell = g_1^\ell g_2^\ell \ldots g_{i-1}^\ell g_{i+1}^\ell \ldots g_r^\ell$$

because the different g_j commute by the result above. Now $g_j^\ell = e$, since $O(g_j)$ divides $|G_j| = p_j^{t_j}$. Thus $x^\ell = e$. Hence order $x \mid \ell$. However, order $x = p_i^{t_i}$ and $(\ell, p_i) = 1$. Thus order $x = 1$. Hence $x = e$. By exercise 5.3 $H \cong G_1 \times G_2 \times \ldots \times G_r$. But then

$$|H| = p_1^{t_1} p_2^{t_2} \ldots p_r^{t_r} = n .$$

Thus $\qquad G = H \cong G_1 \times G_2 \times \ldots \times G_r$ as required.

8. $|G| = 2p$, p odd prime. $(1 + kp)$ subgroups of order p. $(1 + kp) \mid 2p \rightarrow (1 + kp) \mid 2 \rightarrow (1 + kp) = 1$. Thus one Sylow p-subgroup. $(1 + 2k)$ subgroups of order 2. $(1 + 2k) \mid 2p \rightarrow (1 + 2k) \mid p \rightarrow (1 + 2k) = 1$ or p. Let G have one subgroup H of order 2, and one subgroup K of order p. Both are normal by theorem 6.3.4 and cyclic by theorem 3.1.3. $|H \cap K|$ divides 2 and p. Hence $|H \cap K| = 1$ and so $H \cap K = \{e\}$. Let $H = \langle h \rangle$ and $K = \langle k \rangle$. $O(h) = 2$, $O(k) = p$. $[h, k] \in H \cap K = \{e\}$ by same argument as in question 7 above. Thus $hk = kh$. By theorem 2.8.1 (5) $O(hk) = 2p$. Thus $G = \langle (hk) \rangle$ is cyclic.

9. Let $|G| = p^r$. Let $x \in G$. By theorem 3.1.2 $O(x) \mid p^r$. Thus $O(x)$ is a power of p. Conversely let every element of G have order a power of p. Let prime $q \mid |G|$, $q \neq p$. Then there exists $x \in G$ with $O(x) = q$, by theorem 6.3.1. Contradiction. Thus $|G| = p^r$.

10. Let G be non-abelian $|G| = p^3$. Then $|Z(G)| = 1, p, p^2$. By theorem 4.4.2 (3) $|Z(G)| \neq 1$. Suppose $|Z(G)| = p^2$. Then $|G/Z(G)| = p$. Thus $G/Z(G)$ is cyclic and equals

$$\{Z, gZ, g^2Z, g^3Z, \ldots, g^{p-1}Z\} .$$

Hence $\qquad G = Z \cup gZ \cup g^2Z \cup \ldots \cup g^{p-1}Z .$

Let $x, y \in G$. Then $x = g^i c$, $y = g^j d$; $c, d \in Z(G)$. Thus

$$xy = g^i c\, g^j d = g^{i+j} cd = g^j d\, g^i c = yx.$$

Thus G is abelian. Contradiction. Hence $|Z(G)| \neq p^2$. Thus $|Z(G)| = p$ and $|G/Z(G)| = p^2$. By theorem 4.9.1 $G/Z(G)$ is abelian.

11. Let $x \in N_G(U)$. Then $P \subset N$. Hence $x^{-1}Px \subset x^{-1}Nx \subset x^{-1}Ux = U$. Thus P and $x^{-1}Px$ are Sylow p-subgroups of U, hence conjugate in U. Thus for some $y \in U$, $y^{-1}Py = x^{-1}Px$. Thus

$$xy^{-1}P = Pxy^{-1}.$$

Thus $$xy^{-1} \in N_G(P) = N \subset U.$$

Hence $x \in U$. Thus $N_G(U) \subset U$. Always $U \subset N_G(U)$. Thus $U = N_G(U)$.

12. Let $g \in G$. Now $g^{-1}Pg \subset H$ (normal). Thus $g^{-1}Pg$ is a Sylow p-subgroup of H; hence conjugate to P in H by theorem 6.3.3. For some $h \in H$, $g^{-1}Pg = h^{-1}Ph$. Thus $Pgh^{-1} = gh^{-1}P$. Hence $gh^{-1} \in N_G(P)$. Thus $g \in N_G(P).H$. Hence $G \subset N_G(P).H$. Together with $N_G(P)H \subset G$ we conclude that

$$G = N_G(P)H.$$

13. Let $H = \bigcup_i (H \cap P)x_i$, $x_i \in H$ be a right coset decomposition of H relative to its subgroup $H \cap P$. Then $PH = \bigcup_i P(H \cap P)x_i = \bigcup_i Px_i$ is a coset decomposition of PH relative to its subgroup P. Thus

$$[PH : P] = [H : H \cap P].$$

Moreover $p \nmid [PH : P]$. Hence $p \nmid [H : H \cap P]$. Also $H \cap P$ is a subgroup of P, so is a p-group. Thus $H \cap P$ is a Sylow p-subgroup of H.

EXERCISES 7

1. $G = \langle a, b \mid a^3 = b^2 = (ab)^2 = e \rangle$. Then:

$$(ab)^2 = e \to abab = e \to bab = a^2 \to ba = a^2 b.$$

Conversely from $a^3 = b^2 = e$, $ba = a^2b$ we get $a^3 = b^2 = e$, $(ab)^2 = e$.

Thus $G \cong \langle a, b \mid a^3 = b^2 = e, ba = a^2b \rangle \cong S_3$ by section 2.6.

2. $G = \langle x, y \mid x^4 = y^2 = (xy)^2 = e \rangle$. By the given relations:

$$(xy)^2 = e \rightarrow xyxy = e \rightarrow yxy = x^3 \rightarrow yx = x^3y.$$

Conversely from $x^4 = y^2 = e$, $yx = x^3y$ we get

$$x^4 = y^2 = (xy)^2 = e.$$

Thus $\quad G \cong \langle x, y \mid x^4 = y^2 = e, yx = x^3y \rangle \cong D_4$ by section 7.4.

3. $G = \langle x^5 = y^2 = (xy)^2 = e \rangle$. As in questions 1 and 2 above

$$G \cong \langle x^5 = y^2 = e, yx = x^4y \rangle \cong D_5,$$

according to section 7.6.

4. $G = \langle x^6 = y^2 = (xy)^2 = e \rangle \cong \langle x^6 = y^2 = e, yx = x^5y \rangle$ as in questions 1, 2 and 3. This gives a possible group with not more than 12 distinct elements:

$\{e, x, x^2, x^3, x^4, x^5, y, xy, x^2y, x^3y, x^4y, x^5y\}$

which multiply according to the given relations. Take x to be a rotation through 60 degrees in an anti-clockwise sense about an axis through the centre of a regular hexagon perpendicular to its plane. Take y to be a rotation through 180 degrees about an axis L fixed in space perpendicular to a side of the hexagon. Then D_6 has the 12 distinct elements listed above, satisfies the same relations, and has the same multiplication table. Hence $G \cong D_6$.

5. $G = \langle s, t \mid s^3 = t^2 = (st)^3 = e \rangle$.

$ststst = e \rightarrow tstst = s^2 \rightarrow stst = ts^2 \rightarrow tst = s^2ts^2.$

We get not more than 12 distinct elements as follows:

$\{e, s, s^2, t, st, s^2t, ts, ts^2, sts, s^2ts, sts^2, s^2ts^2\}$.

Let M_s denote multiplication of the element by s on the right.
Let M_t denote the same for t. We get:

Element:	e	s	s^2	t	st	s^2t	ts	ts^2	sts	s^2ts	sts^2	s^2ts^2
After M_s:	s	s^2	e	ts	sts	s^2ts	ts^2	t	sts^2	s^2ts^2	st	s^2t
After M_t:	t	st	s^2t	e	s	s^2	s^2ts^2	sts	ts^2	sts^2	s^2ts	ts

These results enable us to obtain the multiplication table, e.g.

$$(s^2ts)(ts^2) = (((s^2ts)t)s)s = ((sts^2)s)s = (st)s = sts .$$

Put $s = (234)$ and $t = (12)(34)$, then A_4 with 12 distinct elements satisfies the relations for G and has the same multiplication table. It follows that we cannot get less than 12 elements from the given relations and in fact $G \cong A_4$.

6. $G = \langle a, b \mid a^7 = b^2 = (ab)^2 = e \rangle$.

As for questions 1, 2, 3, 4,

$$G \cong \langle a, b \mid a^7 = b^2 = e, ba = a^6 b \rangle \cong D_7$$

according to section 7.10.

7. From section 7.6 $D_5' = \langle a \rangle$, $D_5/D_5' \cong C_2$. Sylow subgroups are either order 2 or 5, hence necessarily cyclic by theorem 3.1.3.

From section 7.10, $D_7' = \langle a \rangle$, $D_7/D_7' \cong C_2$. Sylow subgroups are either order 2 or 7, hence cyclic. By section 7.8 $M' = \langle b \rangle$, $M/M' \cong C_4$. Sylow 3-subgroups are order 3 hence cyclic. M has an element of order 4 hence a cyclic Sylow 2-subgroup. All Sylow 2-subgroups are conjugate hence isomorphic. Thus all Sylow 2-subgroups are cyclic.

EXERCISES 8

1. Let f be the natural epimorphism $G \to G/N$. Let f_r be the restriction of f to H. Then

$$f_r(H) = f(H) = \{hN \mid h \in H\} = HN/N .$$

By theorem 4.5.3 $H/\text{Ker}\, f_r \cong HN/N$. By theorem 8.4.1

$$\text{Ker}\, f_r = H \cap \text{Ker}\, f = H \cap N .$$

Thus $\quad H/H \cap N \cong HN/N$.

2. $PN/N \cong P/P \cap N$ by above. Thus PN/N is a p-group. Now

$$[G/N : PN/N] = |G/N|/|PN/N| = \frac{|G|}{|N|} \Big/ \frac{|PN|}{|N|} = |G|/|PN| .$$

But $p \nmid (|G|/|PN|)$ since P is a Sylow p-subgroup of G. Thus

$p \nmid [G/N : PN/N]$. Thus PN/N is a p-group whose index in G/N is not divisible by p. Hence PN/N is a Sylow p-subgroup of G/N.

3. A_n is normal in S_n. Thus $TA_n/A_n \cong T/T \cap A_n$. Now $A_nT \neq A_n$, because T contains odd permutations. Thus $A_nT = S_n$; recall that $[S_n : A_n] = 2$. Hence $S_n/A_n \cong T/T \cap A_n$. Thus $|T/T \cap A_n| = 2$ and so $T \cap A_n$, the subgroup of even permutations in T, is a normal subgroup of index 2 in T. Let $|T \cap A_n| = t$. Now T has m odd permutations. Hence $|T| = t + m$. $|T/T \cap A_n| = 2$. Thus

$$2|T \cap A_n| = |T| = t + m .$$

Thus $2t = t + m$. Hence $t = m$ and $|T| = 2m$.

4. Let $f : G \to S_n$ be the permutation representation of G into the symmetric group on the $n = [G : H]$ cosets of H relative to G as given in section 4.6. $|S_n| = n! = [G : H]!$ $f(G) \not\cong G$, since $|G|$ does **not** divide $|S_n|$ whereas $|f(G)|$ divides $|S_n|$ by Lagrange's theorem. Hence f is **not** one-one. Thus $\operatorname{Ker} f \neq \{e\}$. However, by section 4.6, theorem 4.6.1, $\operatorname{Ker} f \subset H$. Thus $\operatorname{Ker} f$ is the required normal subgroup of G.

5. $|A_5| = 60$. Assume simple. Possible subgroups $\neq A_5$ with orders greater than 12 have orders: 15, 20, 30. Take $G = A_5$ and $|H| = 15, 20, 30,$ in turn in question 4 above. First $|H| = 15$. Then $[A_5 : H] = 4$ and $|A_5| = 60$ does not divide $4! = 24$. Then, by question 4, H contains a normal subgroup of A_5 other than $\{e\}$. Contradiction to A_5 simple. Hence no H of order 15. Next $|H| = 20$. $[A_5 : H] = 3$ and $|A_5| \nmid 3!$. Again contradiction results. Finally $|H| = 30$. Then $[A_5 : H]$ and $|A_5| \nmid 2!$. Again contradiction. Thus A_5 has no subgroup of order greater than 12 other than A_5 itself.

6. Refine both until isomorphic without repetitions by the Schreier-J-H theorem 8.5.1. Now composition series remain the same on refinement without repetitions. Hence the composition series were isomorphic to start with.

7. Refine the compostion series and normal series until isomorphic without repetitions. Then the normal series has become a composition series. Let N be normal in G. Then $\{e\} \subset N \subset G$

is a normal series. By first part of this exercise this can be refined to a composition series which will include N.

8. Infinite cyclic group.

9. $S_3 \supset A_3 \supset \{e\}$ is a solvable series for S_3.

$S_4 \supset A_4 \supset \{(1), (12)(34), (13)(24), (14)(23)\} \supset \{(1)\} = \{e\}$

is a solvable series for S_4.

10. $S_n \supset A_n \supset \{e\}$ is a composition series for S_n because A_n is simple for $n \geq 5$. However, $A_n/\{e\} \cong A_n$ is not abelian. Thus the series is not a solvable series. By theorem 8.6.2, if S_n were solvable then we could refine $S_n \supset A_n \supset \{e\}$ to a solvable series. Since it is a composition series this means that it would have to be already a solvable series. Hence S_n is not solvable for $n \geq 5$. Theorem 8.6.3 could also be used to deduce the required result.

MISCELLANEOUS EXERCISES

1. Class equation gives: $|G| = 1 + h_1 + h_2 + \ldots + h_r$ where 1 refers to class $\{e\}$ and h_t is the number of elements in the tth conjugacy class. Let a_t be an element in the tth class. $N_G(a_t) = \{e, a_t, a_t^{-1}\}$ or $\{e, a_t\}$ when $a_t = a_t^{-1}$. Thus $h_t = |G|/3$ or $|G|/2$. Hence

$$|G| = 1 + m|G|/3 + n|G|/2$$

for some positive integers m and n. Thus $m < 3$, $n < 2$. Take $n = 0$. Possibilities are: $|G| = 1$, $|G| = 1 + |G|/3$, $|G| = 1 + 2|G|/3$. Thus $|G| = 1, 3/2, 3$. Take $n = 1$. Possibilities are: $|G| = 1 + |G|/2$, $|G| = 1 + |G|/3 + |G|/2$, $|G| = 1 + 2|G|/3 + |G|/2$. Thus $|G| = 2, 6, -6$. Altogether the possibilities are: $G = \{e\}$, $|G| = 2, 3, 6$. Thus $G = \{e\}$, C_2, C_3, C_6, S_3. Detailed check for each group shows that: $\{e\}$, C_2, C_3, S_3 satisfy the given conditions.

2. $f: G \to H$. Let $O(a) = n$, $O(f(a)) = m'$. Then

$$(f(a))^n = f(a^n) = f(e) = e .$$

Thus $m' \mid n$. Let f be the natural epimorphism $G \to G/K$. $|G/K| = n$. Let $h \in H$. $f(h) \in G/K$, so by Lagrange's theorem, $O(f(h))$ divides n. Now $O(f(h))$ divides $O(h)$ by above, and $O(h)$ divides $|H| = m$. Thus $O(f(h))$ divides n and m. Now $(n, m) = 1$. Thus $O(f(h)) = 1$. Hence $f(h) = e$ and $h \in \operatorname{Ker} f = K$. Thus $H \subset K$.

3. Define f by $f(g) = \begin{pmatrix} P_1 & P_2 & & P_n \\ gP_1g^{-1} & gP_2g^{-1} & \cdots & gP_ng^{-1} \end{pmatrix}$.

Now gP_ig^{-1} is a Sylow p-subgroup and if $gP_ig^{-1} = gP_jg^{-1}$ then $P_i = P_j$. Thus f is a well-defined function $G \to S_n$. Also $f(gh) = f(g)f(h)$, so f is a homomorphism.

(i) $g \in \operatorname{Ker} f \leftrightarrow f(g)$ is the identity permutation of $\{P_1, P_2, \ldots, P_n\} \leftrightarrow gP_ig^{-1} = P_i$ for all $i \leftrightarrow gP_i = P_ig$ for all $i \leftrightarrow g \in N_G(P_i)$ for all $i \leftrightarrow g \in \bigcap_{i=1}^{n} N_G(P_i)$. Thus $\operatorname{Ker} f = \bigcap_{i=1}^{n} N_G(P_i)$ in general.

(ii) Only one P_i and $N_G(P_i) = G$. Thus $\operatorname{Ker} f = G$.

(iii) Take $G = A_4$, $p = 3$. $|A_4| = 12 = 2^2 \times 3$. Sylow 3-subgroups P_i are order 3 hence cyclic. Section 7.8 gives 4 Sylow 3-subgroups of A_4. Thus $4 = [A_4 : N_{A_4}(P_i)]$. Thus $N_{A_4}(P_i)$ is cyclic order 3. Now $N_{A_4}(P_i) \supset P_i$. Thus $N_{A_4}(P_i) = P_i$. Hence $\operatorname{Ker} f = \bigcap_{i=1}^{4} P_i = \{e\}$.

Hence f is a monomorphism $A_4 \to S_4$.

4. By exercise 4.12 $G/Z(G) \cong I(G)$. $|I(G)| = 1 \leftrightarrow Z(G) = G \leftrightarrow G$ is abelian. Elements (12), (13) of order 2 generate S_3. Thus automorphism f is determined by its effect on these elements. Also f must send any element of order 2 onto an element of order 2. There are 3 elements of order 2. Thus there are at most $3! = 6$ automorphisms of S_3. Now $I(S_3) \cong S_3/Z(S_3) = S_3$, because $Z(S_3) = \{(1)\}$. Now $|S_3| = 6$. Hence $A(S_3) = I(S_3) \cong S_3$.

5. (i) Let $n = sd$. Put $|H| = \langle sg \rangle$ then $|H| = d$. Let $h \in H$ then $O(h) \mid |H|$. Thus $O(h) \mid d$. Hence $H \subset G_d$. Let $x \in G_d$. Then $O(x) \mid d$. Let $O(x) = t$. Then $d = rt$. Thus $n = srt$. Now $x \in G$,

so $x = ug$ for some $0 \leqq u < n$. Also $0 = tx = tug$. Thus $n \mid tu$. Hence $tu = vn = vsrt$ for some v. Thus $u = vsr$. Hence

$$x = vsrg = vr(sg) = vr(sg) \in H .$$

Thus $G_d \subset H$. Altogether $H = G_d$.

(ii) t_g generates $G = \langle g \rangle \leftrightarrow tg$ and g have the same order $\leftrightarrow 1 = (t, n)$ by theorem 2.8.1.

(iii) $f(x) = tx$, $(t, n) = 1$, defines an automorphism of $G = \langle g \rangle$. Conversely let f be an automorphism of G. Then $f(g)$ must generate G. Thus $f(g) = tg$, $(t, n) = 1$ by (ii) above. Let $x \in G$. Then $x = ug$ and $f(x) = uf(g) = utg = tx$. Thus $f(x) = tx$ for all $x \in G$, $(t, n) = 1$.

6. Take $A = \langle a \rangle$, $B = \langle b, c \rangle$. Then A is cyclic order 4 and B is a Klein 4-group. Let $x \in A \cap B$. Then $x \in \{e, a, a^2, a^3\}$ and $x \in \{e, b, c, bc\}$. However, b, c, bc cannot be expressed in terms of a from the given relations. Thus $x = e$. Then

$$|AB| = |A||B|/|A \cap B| = 16.$$

Thus $AB = G$.

7. $H \cong g^{-1}Hg$ have the same order. However, there is only one subgroup of given order of a finite cyclic group K. Thus $H = g^{-1}Hg$.

8. $H \subset K$. Thus $g^{-1}Hg \subset g^{-1}Kg = K$ for all $g \in G$. By question 7 above $H = g^{-1}Hg$ for all $g \in G$. Hence H is normal in G.

9. Multiplication is well-defined because if $m_1, m_2 \in G$ then $(m_1, n) = 1 = (m_2, n)$ so $(m_1m_2, n) = 1$. Hence $m_1m_2 \pmod{n} \in G$. Associativity follows from that for multiplication of integers. The identity is 1. Let $m \in G$. Because $(m, n) = 1$ we have $Mm + Nn = 1$ (see [G]). Thus $Mm = 1 \pmod{n}$ and $(M, n) = 1$. Thus $M \in G$ is the inverse of m in G. Hence G is a group, abelian since multiplication of integers is commutative. Take $n = 16$. $G = \{1, 3, 5, 7, 9, 11, 13, 15\}$.

Element:	1	3	5	7	9	11	13	15
Order:	1	4	4	2	2	4	4	2

G is abelian order 8. Possibilities are: C_8, $C_2 \times C_4$, $C_2 \times C_2 \times C_2$. Now C_8 has an element order 8 and $C_2 \times C_2 \times C_2$ has all elements order 2. Thus $G \cong C_2 \times C_4$. Let m be odd. Then $m \equiv g \pmod{16}$ for some $g \in G$. Thus 16 divides $m^4 - 1 \leftrightarrow m^4 \equiv 1$ modulo $16 \leftrightarrow g^4 = 1$ in the group G above, where $m \equiv g \pmod{16}$. Now the orders of the elements in G are 1, 2, 4. Thus $g^4 = 1$ for **all** $g \in G$. Hence 16 divides $m^4 - 1$ for any odd m.

10. G non-abelian. Let $G/Z(G)$ be cyclic. Then

$$G/Z(G) = \{Z(G),\ gZ(G),\ \ldots,\ g^t Z(G),\ \ldots\}.$$

Thus
$$G = Z(G) \cup gZ(G) \cup \ldots \cup g^t Z(G) \cup \ldots .$$

Let $x, y \in G$. Then $x = g^i c$, $y = g^j d$; $c, d \in Z(G)$. Thus

$$xy = g^{i+j} cd = yx.$$

Hence G is abelian. Contradiction. Hence $G/Z(G)$ is not cyclic. Let $A(G)$ be cyclic. Then $I(G) \subset A(G)$ is cyclic. Now $I(G) \cong G/Z(G)$. Hence $G/Z(G)$ is cyclic. Contradiction. Hence $A(G)$ is not cyclic.

11. Let finite G have t subgroups other than $\{e\}$ and G.

(i) $t \leqq 2$. Let G be non-cyclic. Then G has at least 2 generators a, b. $\langle a \rangle$ and $\langle b \rangle$ are subgroups $\neq \{e\}$ or G. Thus $\langle (ab) \rangle = \langle a \rangle$ or $\langle b \rangle$. If $a \in \langle (ab) \rangle$ then $b = a^{-1}(ab) \in \langle (ab) \rangle$. If $b \in \langle (ab) \rangle$ then $a \in \langle (ab) \rangle$. Hence if $G = \langle a, b \rangle$ then $G = \langle (ab) \rangle$. Contradiction. If $G = \langle a, b, c, \ldots \rangle$ with no redundant generators, then $\langle a \rangle, \langle b \rangle, \langle c \rangle, \ldots$ are distinct subgroups. Thus if $t \leqq 2$ then G must be cyclic.

(ii) Let $t = 0$. By (i) G is cyclic. Since $t = 0$, G is simple. By exercise 4.20 G is cyclic of prime order.

(iii) Let $t = 1$. Then G is cyclic by (i). By theorem 3.5.1 $|G|$ has divisors 1, $|G|$, and some $1 < d < |G|$. If $|G|$ has prime factors p, q then $|G|$ has divisors 1, p, q, $|G|$. Hence $|G| = p^r$ with divisors 1, p, p^2, ..., p^r. Thus $r = 2$ and $|G| = p^2$.

(iv) Let $t = 2$. By (i) G is cyclic. If $|G| = p^r$ then $r = 3$, to give 4 divisors 1, p, p^2, p^3. If $|G| = pqu \ldots$ then divisors are 1, p, q, u, pqu, ... at least. Thus $|G|$ must involve only 2 distinct primes p, q. If $|G| = p^{s_1} q^{s_2}$; $s_1 > 1$, $s_2 > 1$

then there are more than 4 divisors. Hence more than 4 subgroups and $t > 2$. Thus $|G| = pq$. Then divisors are 1, p, q, pq giving $t = 2$ as required. Thus with $t = 2$, $|G| = p^3$ or pq.

12. Let $K = \bigcup_{i=1}^{t} k_i H$, $G = \bigcup_{j=1}^{s} g_j K$. Then $G = \bigcup g_j k_i H$, where the union is disjoint over the cosets $g_j k_i H$; ($i = 1, 2, \ldots, t$ and $j = 1, 2, \ldots, s$). Thus

$$[G : H] = st = [G : K][K : H].$$

$H_1 = \bigcup_{i=1}^{r} x_i (H_1 \cap H_2)$ is left coset decomposition of H_1 relative to $H_1 \cap H_2$. Then $H_1 H_2 = \bigcup_{i=1}^{r} x_i (H_1 \cap H_2) H_2 = \bigcup_{i=1}^{r} x_i H_2$ is also disjoint (check). $G \supset H_1 H_2 \supset H_2$. Thus

$$r = [H_1 : H_1 \cap H_2] = [H_1 H_2 : H_2] \leq [G : H_2] = 3.$$

Similarly $[H_2 : H_1 \cap H_2] \leq 3$. Thus $[H_1 : H_1 \cap H_2] = 1, 2$, or 3. If $[H_1 : H_1 \cap H_2] = 1$ then $H_1 = H_1 \cap H_2$. Thus $H_1 \subset H_2$. But $|H_1| = |H_2| = |G|/3$. Hence $H_1 = H_2$. Contradiction. Thus

$$[H_1 : H_1 \cap H_2] = 2 \text{ or } 3.$$

Now by the first part of this question:

$$[G : H_1 \cap H_2] = [G : H_1][H_1 : H_1 \cap H_2] = 3[H_1 : H_1 \cap H_2].$$

Hence $[G : H_1 \cap H_2] = 6$ or 9.

Let $|G| = 51 = 3 \times 17$. Since $17 \mid |G|$ there exists $g \in G$ with $O(g) = 17$, by theorem 6.3.1. Let $H = \langle g \rangle$, $|H| = 17$. Let L be any other subgroup of G of order 17. $[G : H] = [G : L] = 3$. By above $[G : H \cap L] = 6$ or 9. Now $|H \cap L| = 1$ or 17 by Lagrange's theorem. Hence if $H \neq L$, then $H \cap L = \{e\}$. Thus $[G : H \cap L] = 51$. Contradiction. Thus $H = L$. Let $g \in G$. Then $g^{-1}Hg \cong H$. Hence $|g^{-1}Hg| = 17$. By the above $g^{-1}Hg = H$. Thus H is normal in G. (This follows also from theorem 6.3.4.)

Now 3 divides 51. Hence G has a cyclic subgroup K of order 3. $H \cap K = \{e\}$ by Lagrange's theorem. Because H is normal, $HK = KH$ is a subgroup of G and $|HK| = |H||K|/|H \cap K| = 51$ by theorem 3.4.2. Thus $G = HK$. By theorem 6.3.4 K is also normal in G.

By theorem 3.1.3 $H = \langle h \rangle$, $K = \langle k \rangle$, $[h, k] \in H \cap K = \{e\}$. Thus $hk = kh$. By theorem 2.8.1(5) $O(hk) = 51$. Thus $G = \langle hk \rangle$ is cyclic.

13. $|G| = 2$ is trivial. Let $|G| > 2$. Let $a \neq b \in G$. Then $a^2 = b^2 = (ab)^2 = e \to abab = e \to bab = a \to ab = ba$. Hence G is abelian. By exercise 6.9 $|G| = 2^r$. By the structure theory for abelian groups, with multiplicative notation, G is isomorphic to a direct product of cyclic groups C_{2^s}. Now C_{2^s} has elements of order >2 unless $s = 1$. Thus $G \cong$ direct product of cyclic groups of order 2. Let $|G| = 4$. By theorem 3.1.2 the elements of G have orders 1, 2, or 4. If $O(g) = 4$ then $G = \langle g \rangle$ is cyclic. Otherwise all elements $\neq e$ of G are order 2. Then first part shows that:

$$G \cong C_2 \times C_2 \cong \text{Klein 4-group } K_4.$$

14. Let $g \in G$. By theorem 3.1.2 $O(g)$ divides $|G| = n$. Thus $n = O(g)r$. Hence

$$g^n = g^{O(g)r} = e .$$

Let $g \in G = \{1, 2, \ldots, p-1\}$. By above $g^{p-1} = 1$. Hence $g^{p-1} \equiv 1 (\text{mod } p)$. Let $a \in \mathbf{Z}$. Then $a \equiv g (\text{mod } p)$ for some $g \in G$. Thus $a^{p-1} \equiv 1 (\text{mod } p)$. Hence $a^p \equiv a (\text{mod } p)$.

15. $|G| = 24 = 2^3 \times 3$. By Sylow's third theorem G has 1 or 3 Sylow 2-subgroups of order 8 and 1 or 4 Sylow 3-subgroups of order 3.

(i) By theorem 3.5.1 C_{24} has 1 subgroup of order 3 and 1 subgroup of order 8.

(ii) S_4.$\{(1), (123), (132)\}$ and $\{(1), (124), (142)\}$ are 2 distinct subgroups of order 3. Hence there are 4 subgroups of order 3 in S_4. By exercise 4.18 no subgroup of S_4 of order 8 is normal. Hence S_4 has 3 subgroups of order 8.

(iii) $A_4 \times C_2$. $\{(1), (123), (132)\} \times \{e\}$ and $\{(1), (124), (142)\} \times \{e\}$ are subgroups of order 3. Hence $A_4 \times C_2$ has 4 subgroups of order 3. By exercise 4.18

$$K = \{(1), (12)(34), (13)(24), (14)(23)\}$$

is a normal subgroup of A_4. Thus $K \times C_2$ is a normal subgroup of order 8 of $A_4 \times C_2$. Hence $A_4 \times C_2$ has 1 subgroup of order 8.

FURTHER READING AND REFERENCES

INTERMEDIATE (Advanced Undergraduate; First Year Postgraduate)

1. Johnson, D. L. *Presentation of Groups*, C.U.P., London (1976). 2nd edition: *Topics in the Theory of Group Presentations*, C.U.P., London (1980).
(This is an excellent introduction to combinatorial group theory, i.e. generators and relations.)

2. Macdonald, I. D. *The Theory of Groups*, O.U.P., London (1968).
(A good introduction to nilpotent and solvable groups.)

3. Rose, J. S. *A Course on Group Theory*, C.U.P., London (1978).
(Special feature is the treatment of group actions.)

4. Rotman, J. J. *The Theory of Groups*, Allyn and Bacon, Boston (1973).
(Particularly good on homological algebra, generators, relations and word problems.)

5. Weinstein, M. *Examples of Groups*, Polygonal Publishing House, Passaic, New Jersey, U.S.A. (1977).
(Excellent for group constructions such as direct products, semi-direct products, wreath products, etc.)

ADVANCED

6. Burnside, W. *Theory of Groups of Finite Order*, Dover, New York (1955).
(First published 1911. Old-fashioned notation, but still rewarding reading.)

7. Hall, M. *Theory of Groups*, Chelsea Publishing Co., New York (1976).

8. Huppert, B. *Endliche Gruppen, Vol. 1*, Springer-Verlag, Berlin (1967). Vol. 2 is expected shortly.
(Comprehensive and lucid.)

9. Kurosh, A. G. *Theory of Groups, Vols 1, 2, 3*, Chelsea Publishing Co., New York.
(Comprehensive but largely devoted to infinite groups.)

10. Lyndon and Schupp, *Combinatorial Group Theory*, Springer-Verlag, Berlin (1977).

11. Magnus, Karrass, and Solitar, *Combinatorial Group Theory*, Dover, New York (1976).

12. Neumann, H. *Varieties of Groups*, Springer-Verlag, Berlin (1967)

13. Robinson, D. J. S. *Finiteness Conditions and Generalised Soluble Groups, 1 and 2*, Springer-Verlag, Berlin (1972).

14. Wielandt, H. *Finite Permutation Groups*, Academic Press, New York (1964).

SOURCES OF FURTHER PROBLEMS

15. Dixon, J. D. *Problems in Group Theory*, Blaisdell Publishing Co., Waltham (1967).

16. Lyapin and others, *Exercises in Group Theory*, Wolters-Noordhoff Publishing, Groningen (1972).

REFERENCES

17. Carter, R. W. *Simple Groups of Lie Type,* John Wiley and Sons, London (1972).

18. Coxeter and Moser, *Generators and Relations for Discrete Groups,* Springer-Verlag, Berlin (1972). (Fourth edition, 1980.)

19. Dornhoff, L. *Group Representation Theory, A and B,* Marcel Dekker, New York (1971-72).

20. Escher, M. C. *The Graphic Work of M. C. Escher,* Pan/Ballantine, London (1973).

21. Feit, W. and Thompson, J. G., 'Solvability of Groups of Odd Order', *Pacific Journal of Mathematics,* **13**, pp. 775-1029 (1963).

22. Gruenberg, K. W. *Cohomological Topics in Group Theory,* Springer-Verlag, Berlin (1970).

23. Hall, M. and Senior, J. K. *The Groups of Order* 2^n, $(n \leq 6)$, Macmillan, New York (1964).

24. Hall, P. 'Classification of Prime Power Groups', *J. für die reine u.ang.Math.,* **182**, pp. 130-41 (1940).

25. Ledermann, W. *Introduction to Group Characters,* C.U.P., London (1977).

26. MacLane, S. *Homology,* Springer-Verlag, Berlin (1963).

27. Powell and Higman, *Finite Simple Groups,* Academic Press, London (1971).

28. Stewart, I. *Galois Theory,* Chapman and Hall (1973).

29. Sylow, L. 'Théorèmes sur les groupes de substitutions', *Math. Ann.,* **5**, pp. 584-94 (1872).

FURTHER READING AND REFERENCES FOR SCIENTISTS

30. Leech and Newman, *How to Use Groups,* Methuen (1969).

31. Miller, W. *Symmetry Groups and Their Applications,* Academic Press, New York (1972).

32. Wenninger, M. J. *Polyhedron Models*, C.U.P., London (1971).

33. Weyl, H. *Symmetry*, Princeton University Press, Princeton (1952)

34. Yale, P. B. *Geometry and Symmetry*, Holden-Day, San Francisco (1968).

GENERAL REFERENCE

[G] Gardiner, C. F. *Modern Algebra; A Natural Approach, with Applications*, Ellis Horwood Ltd., Chichester; Halsted Press, John Wiley and Sons, New York (1981).

INDEX